AF712054

ENVIRONMENTAL ACTION

Analyze **C**onsider options **T**ake action **I**n **O**ur **N**eighborhoods

CHEMICALS
Choosing Wisely

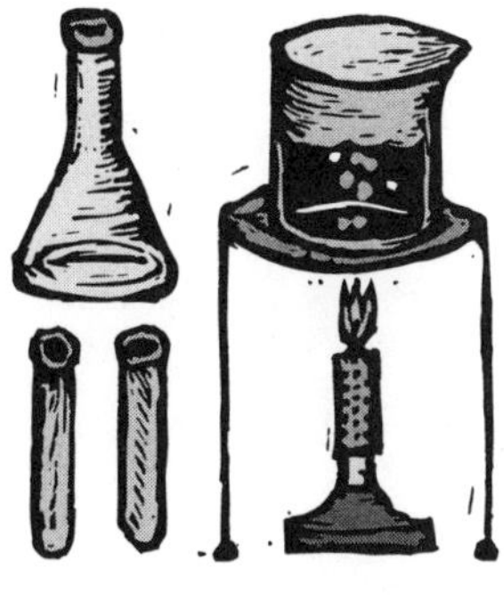

KA-BOOM

ENVIRONMENTAL ACTION

Analyze Consider options Take action In Our Neighborhoods

CHEMICALS
Choosing Wisely

A Student Audit of Resource Use

TEACHER RESOURCE GUIDE

E2: ENVIRONMENT & EDUCATION

DALE SEYMOUR PUBLICATIONS®
MENLO PARK, CALIFORNIA

Developed by E2: Environment & Education™, an activity of the Tides Center.

Managing Editor: Catherine Anderson
Senior Editor: Jeri Hayes
Senior Production Coordinator: Alan Noyes
Production/Manufacturing Director: Janet Yearian
Design Manager: Jeff Kelly
Text and Cover Design: Lynda Banks Design
Art: Rachel Gage, Andrea Reider
Composition: Andrea Reider
Clip Art Illustrations: Copyright © Art Parts, Courtesy Art Parts, 714-834-9166

This book is published by Dale Seymour Publications®,
an imprint of Addison Wesley Longman, Inc.

Dale Seymour Publications
2725 Sand Hill Road
Menlo Park, CA 94025
Customer Service: 800 872-1100

Copyright © 1998 The Tides Center/E2: Environment & Education
All rights reserved.
Printed in the United States of America.

Limited reproduction permission: The publisher grants permission to individual teachers who have purchased this book to reproduce the blackline masters as needed for use with their own students. Reproduction for an entire school or school district or for commercial use is prohibited.

DALE SEYMOUR PUBLICATIONS®

Printed on acid-free,
85% recycled paper
(15% post-consumer).
Printed using soy-based ink.

ISBN 0-201-49534-1
DS36864
2 3 4 5 6 7 8 9 10-ML-01 00

ACKNOWLEDGMENTS

Tom and Rampa Hormel, Christopher and Luanne Hormel, Angela Hormel Ocone, and Diane Ives from the Global Environment Project Institute (GEPI) for their unwavering support in developing this action program for students

Jay Dean Paschall of Global Learning and Observation to Benefit the Environment (GLOBE) for developing the concept for this program (originally called EarthTime)

Human-i-Tees for generous support of the review and pilot school portion of the program

The Center for Environmental Education for use of their resource materials

Leslie Crawford and Amy E. Holm for writing and producing *Environmental ACTION*

EarthTime Contributors

Larry J. Barnes, Ketchum, ID
Betty Bell, Ketchum, ID
J. B. Burrell, Ketchum ID
Scott Graves, T.R.E.E., Boise, ID
Irene S. Healy, Wood River High School, Hailey, ID
Carrie Hislaire, Sun Valley, ID
Michelle Richer, Ketchum, ID
Doug Wilson, Ketchum, ID

Environmental ACTION Contributors

Martin J. Byhower, Chadwick School, Palos Verdes, CA
Jennifer Daza, E2: Environment & Education
Bruce Harlan, St. Matthew's Parish School, Pacific Palisades, CA
Anne Kirstin Holm, student at Yale University
Laura Jean Moore, San Diego, CA
Bonnie Slagel, Los Angeles, CA

Pilot School Participants

Michele Cheyne, Hamilton High School, Milwaukee, WI
Neil Coen, East Magnet School, Kansas City, MO
Steve Engleman, Paul Revere Middle School, Los Angeles, CA
Leroy Hickerson, Alcee Fortier Senior High School, New Orleans, LA
Stephen Hopkins, Sly Park Environmental Education Center, Pollack Pines, CA
Dick Jordan, Boise High School, Boise, ID
Donna Gross McDaniel, Alcee Fortier Senior High School, New Orleans, LA
Dennis Pilien, Bravo Medical Magnet, Los Angeles, CA
Scott Sala, Cory Elementary School, Denver, CO
Pat Shepard, Miller-South School for Visual and Performing Arts, Akron, OH

Educational Reviewers

Karl Abrahms, Saddleback College, Mission Viejo, CA
Tammy Bird, Crenshaw High School, Los Angeles, CA
Alison L. Brown, Beekmantown Central School, Plattsburgh, NY
Martin Byhower, Chadwick School, Palos Verdes, CA
Joan Grimm, Department of Environmental Quality, Portland, OR
Bruce Harlan, St. Matthew's Parish School, Pacific Palisades, CA
Kay Ice, Educational Development Specialists, Lakewood, CA
Kurt Leuschner, Mira Costa High School, Manhattan Beach, CA
Stephanie Wald, Allan Hancock College, Santa Maria, CA

Professional Reviewers

Christopher Balthasar, EarthSave Foundation, Santa Monica, CA
Mari Clements, Nutrition Information Center, Crozer-Chester Medical Center, Upland, PA
Wayne R. Gould, Southern California Edison, Rosemead, CA
Joe Haworth, Sanitation Districts of Los Angeles County, Whittier, CA
Richard Heede, Rocky Mountain Institute, Snowmass, CO
Steven Hulbert, Saturn of Olympia, Olympia, WA
Sherman Morrison, North American Coalition on Religion and Ecology, Washington, DC
Gary Petersen, Recycle America, Los Angeles, CA

E2 Board of Advisors

Peter Corcoran, Bates College
Irene Healy, Wood River High School
Christopher Hormel, Global Environment Project Institute (GEPI)
Rampa Hormel, GEPI
Tom Hormel, GEPI
Dean Paschall, Global Learning and Observation to Benefit the Environment (GLOBE)
Drummond Pike, Tides Foundation and Tides Center

CONTENTS

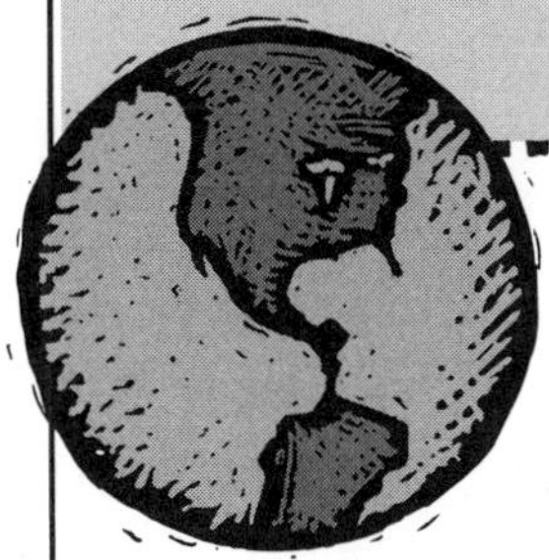

Welcome to Environmental ACTION!

EXPLORE the Issues

ANALYZE

CONSIDER OPTIONS

TAKE ACTION

Appendices

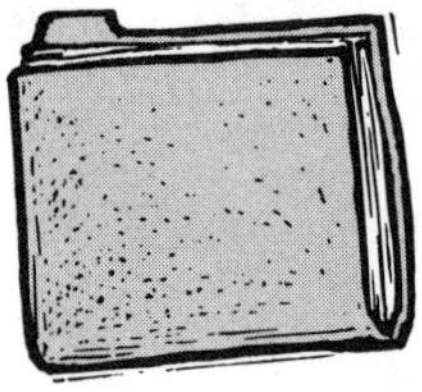

Blackline Masters

Welcome to Environmental ACTION!

About the Environmental ACTION Program

As the natural resources crisis reaches global proportions—air, water, and land pollution; limited supplies of fossil fuels; threatened and endangered habitats; species facing extinction—people must acknowledge the enormous role that they have played in creating or exacerbating the problems and the enormous role they can potentially play to alter the course of events. As educators, we feel it is vital to give students the information and skills they will need in order to take action. We need to help students learn to look critically at environmental issues and take personal responsibility for finding solutions by asking questions: What is the problem? What causes the problem? What impact does behavior have on the problem? What changes can be made or actions taken to help? By introducing an action-based curriculum, we not only begin preparing the next generation for dealing with difficult issues, we teach students how to live healthier lives in the process.

The school is an ideal laboratory for this hands-on experiment because it provides a real-world model. Students gather information about the school, analyze environmental issues within the context of the information gathered, determine positive alternatives, and practice implementing solutions right in the school setting. For example, students may decide that by using energy-efficient lighting, implementing water conservation programs, moving toward environment-friendly landscaping, or using nontoxic chemicals to clean and maintain facilities, they can improve the school environment, reduce resource use, and perhaps even save money. They present their proposal for changes to the school for approval. If the changes are implemented, the school then becomes a useful living paradigm; students can participate in and observe the changes, monitor results, and extend their knowledge to their own homes. It is this learning/doing, school/community partnership that makes Environmental ACTION a unique environmental education curriculum.

Program Mission and Objectives

Environmental ACTION was developed in response to the need for environmental education materials that emphasize personal responsibility and positive action. The program's mission is to empower students with the knowledge and skills necessary to make meaningful changes that can be carried into the future. Many educators have been challenged to find appropriate materials and the necessary teaching support to offer this kind of environmental education to their students. Environmental ACTION meets these needs by providing a relevant, supportive, clearly structured curriculum specifically aimed at middle and secondary students.

The six modules in the program provide step-by-step instruction on how to investigate real-world environmental issues and present opportunities to learn and practice action skills in the context of these issues. By creating a laboratory within the school community, the program gives students the opportunity to learn and develop personal responsibility through practical application.

The curriculum has the following objectives:

- to promote awareness of environmental issues through real-world investigations
- to build the knowledge and skills needed to analyze, investigate, and offer solutions to environmental problems

- to encourage practical application of knowledge and skills in issue resolution
- to assist students in becoming responsible citizens by involving them in the extended school community

Program Description

The program consists of six modules designed for use in middle and secondary schools. Each module includes a Teacher Resource Guide and Student Edition. Following is a brief description of the six modules:

Energy Conservation
Students explore the sources, production, uses, and environmental effects of energy. They apply their learning by examining ways to improve the energy efficiency of their school and homes.

Food Choices
Students investigate the effects of food production, diet, and nutrition on human health and the environment. Students analyze the school's food service program to identify healthy choices and practices.

Habitat and Biodiversity
Students study the importance of biological diversity, landscape management, xeriscaping, composting, and integrated pest management (IPM). Using the school as a research laboratory, students assess the current landscaping, then evaluate its present health and environmental impact. This module also contains a step-by-step guide to creating an organic garden and seed bank.

Chemicals: Choosing Wisely
Students investigate the types of materials, chemical products, cleaning supplies, and pesticides used in their school—how they are used, stored, and disposed of, and their potential effects on human health and the environment. Students develop a plan for implementing earth- and human-friendly alternatives for school and home.

Waste Reduction
Students sort and analyze school garbage to identify recyclable and compostable materials. They formulate a plan to reduce their consumption and waste at school and at home, including developing a recycling program or improving an existing one.

Water Conservation
After an introduction to water consumption and quality issues, students conduct an audit of water usage and efficiency on the school campus. Using the school as a research laboratory, students develop strategies for implementing water conservation at school and at home.

Options for Using the Program

These six modules are part of a complete yet flexible curriculum package. Because they are designed to be used either in conjunction with one another or as stand-alone units, a variety of teaching options are offered. For example, one module can be taught year after year, different modules can be taught in consecutive years, or different modules can be taught in a given year. Modules may be taught by a team, or a group of teachers may choose to simultaneously present different modules.

Using the Program Outside School

In response to a growing number of requests, Environmental ACTION has been designed for use by groups that operate outside a formal school setting. Issue investigation and action activities in the Teacher Resource Guide and Student Edition can be easily adapted to fit different venues and circumstances. For example, an ecology club might use the program at a school, community center, or local business; church groups can use it to explore their stewardship responsibilities within the church community or as an outreach project; community-based organizations can conduct the activities at their centers as part of their efforts to improve the quality of life within the local community. Wherever choices are made with respect to natural resource consumption, there exists an opportunity for investigation, evaluation, and action—and Environmental ACTION can provide a framework for these efforts.

Program Components

Each module consists of a Student Edition and a Teacher Resource Guide that complement one another and provide maximum flexibility for teachers.

Student Edition

Each activity is presented using a clear and consistent format that puts students in charge of their own learning. Setting the Stage offers discussion questions that will pique students' interest and direct their attention to the objectives of the lesson. Vocabulary is included as appropriate. The Focus part of the activity provides the core of the lesson, leading students through a step-by-step process. Central to this section is an activity sheet, which can be duplicated for students to complete individually or in groups. (Blackline masters for all activity sheets are located at the back of the Teacher Resource Guide.) It's a Wrap reviews student work and provides closure to the activity. The Home activity allows students to apply their learning outside the school environment.

Teacher Resource Guide

The material in the Teacher Resource Guide duplicates key text in the Student Edition—questions, assignments, and activity sheets. Concepts and objectives are itemized, special vocabulary is listed, and student responses are suggested. Activity sheets in the Teacher Resource Guide include annotated answers. The Teacher Resource Guide contains the same Issues and Information section and Glossary that appear in the Student Edition, along with some additional resources and assessment tools. The blackline masters for all activity sheets are located at the end of the Teacher Resource Guide.

Moving Through a Module

Each activity within a module is a separate lesson, and most are designed to take only one class period. The program consists of four types of activities: Explore the Issues, Analyze, Consider Options, and Take Action. The teaching methodology, illustrated below, presents activities in a sequence that first introduces students to global environmental issues, then requires them to apply critical thinking skills to environmental issues in their immediate environment, and finally encourages them to take responsibility through independent action.

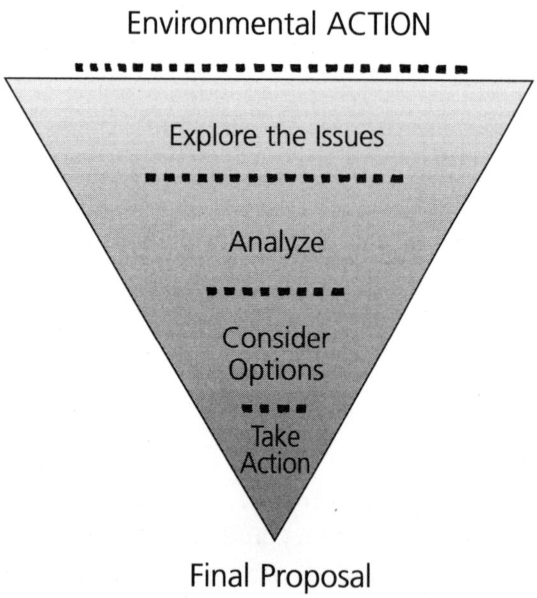

Explore the Issues Activities

In the first four activities in each module, students are given an overview of the environmental topics that pertain to the issues they will focus on in their investigation of school, home, and community. The Explore the Issues activities gives students opportunities to gather information and build environmental awareness. Students will work individually, in small groups, and as a class.

ACT Activities

The remaining activities are structured to provide an environment for students to move from awareness to choice to action. The three-part teaching sequence—ACT—reflects the following process:

- Analyze
- Consider Options
- Take Action

Analyze Activities Students focus their attention on the school campus or other facility. Working in small groups, students develop and implement a plan for collecting and recording data. Finally, students identify environmental, financial, and human health costs and benefits of current school practices affecting, for example, energy use, waste disposal, or whatever topic is the focus of the particular module.

Consider Options Activities Students investigate alternative practices, services, technologies, and products. Then they evaluate each on a costs/benefits basis, using material in the Issues and Information section at the back of the book.

Take Action Activities Based on the options that students have discovered, groups of students recommend changes in practices and products. The class then discusses and evaluates the recommendations and ultimately reaches a consensus as to which recommendations to propose to the school administration or environmental committee. (Note: See Community Recognition, p. 9, for further discussion on local outreach.) The closing activities involve writing the proposal. This final step in each Environmental ACTION module is intended to launch students on a course for creating a healthier and more balanced environment.

Home

Each activity includes ideas for parallel investigations and analyses that students can do at home. These activities reinforce the learning going on at school and provide an opportunity for students to apply the knowledge to practices and choices made at home. Family members may also choose to get involved. The Home activities are optional; they are not required for the completion of the school environmental audit or other activities within the module. We suggest that students' Home assignments be written in a Journal. (See Student Journals, below.)

Act Locally

In order to effect meaningful change in areas related to human health and the environment, it is essential that students practice taking action. As with other areas of learning, practicing a behavior makes it grow into habit. Therefore, Act Locally activities—straightforward outreach projects to present to the school or community—are provided for use after the Analyze and Consider Options sections. Students can choose to implement one of the suggested activities or develop an activity of their own. In either case, these Act Locally activities serve as opportunities for students to practice putting their ideas into action.

Student Journals

Student work on the Home assignments at the end of each activity can be recorded in student Journals, along with notes, sketches, data collection, and other student writing that takes place during the module. Journals may be kept private and used for referral and note-taking during discussions, or they may be used to record answers to specific questions in the Focus or It's a Wrap sections. Since some Journal material may be turned in and some kept for personal use only, a loose-leaf notebook or folder is an ideal choice.

You may want to extend the use of the Journal by suggesting additional writing assignments that call on students to make observations about what they have noted, synthesize or analyze the data, compare and contrast information, draw conclusions, or make generalizations. All of these activities provide opportunities for students to use critical thinking skills and communicate ideas—valuable experiences for most middle and high school students. Additional ideas for Journal assignments can be found in the Cross-Curricular Activities Chart for the module. (See Cross-Curricular Suggestions, below.)

Issues and Information

An Issues and Information appendix appears in both the Student Edition and the Teacher Resource Guide. This appendix contains a variety of materials students need to complete the activities and develop a deeper understanding of the issues they are studying. The appendix is divided into easily accessible sections and, in some modules, may be accompanied by illustrations.

Cross-Curricular Suggestions

At the end of the introductory material for each module, you will find specific suggestions for cross-curricular activities and extensions in the following content areas: history/social studies, language arts, math, and science. These suggestions offer opportunities to extend study activities to higher cognitive skill levels and/or to meet your school's curriculum framework with activities that link directly to specific components.

Teaching Environmental ACTION

Scheduling and Managing Modules Start a module with enough time—18–20 class sessions—for students to conduct research and implement changes. Since students will be needing activity sheets for all the activities, you may wish to photocopy them before starting a module. The blackline masters are located at the end of the Teacher Resource Guide. Before starting the first activity, introduce students to the project, administer the Content Quiz and Student Survey provided in the Blackline Masters/Assessment Tools section, and familiarize students with the Student Edition. You may also wish to have students write to agencies, organizations, manufacturers, and businesses (listed in the Teacher Resources section of the Teacher Resource Guide) in order to obtain materials and request information that can be used as the module progresses. Once the students are working in their Action Groups to investigate and analyze their subjects, additional research time outside of the actual class sessions may need to be scheduled.

Skipping or Modifying Activities Since all of the activities have been carefully designed to help students build their investigation of the school environment, we encourage you to complete each activity. If you do omit or modify an activity, be sure that students incorporate the activity's objectives into the project. You may be able to teach or facilitate your students' comprehension of a learning objective in a different way or combine activities in order to make sure that your students do not miss meaningful steps in the discovery process.

The Teacher's Role The Environmental ACTION modules are activity driven. The scope and sequence has been designed carefully so that the transition from teacher- to student-directed activities happens gradually and within well-outlined parameters. In the beginning, you will be in the more traditional role of the teacher, providing information and leading the students through the Explore the Issues activities. As the class begins the ACT activities, you will assume the role of facilitator for your students' efforts.

Small group work will be the activity most likely to affect your role in relation to the students. Although you will continue to supervise them, you will not be directing their work. Students working in Action Groups will be responsible for keeping on task, catching their mistakes, supporting each other, and meeting their goals. Students may not always carry out their work in a way that you would have directed, but this learning process allows them to make mistakes and correct their actions.

As your students work through the actual audit and your role becomes more that of a facilitator, your responsibilities may include some or all of the following:

- serving as liaison between the students and the school administration
- keeping the students' work on schedule
- monitoring the groups' interactions
- advising Action Groups about procedures and resources and assisting with problem-solving
- reserving reference and resource books in the school library or classroom for student research

Toward the end of the ACT sequence, students will meet again as a class to finalize their proposal. At this point, you will evaluate the group dynamics of your students and assess the degree of facilitation necessary. After students write the proposal and present it to the school administration or environmental committee, the actual implementation steps will be determined by the school's willingness to take action and your students' commitment to the project.

Student Project Coordinator To provide more leadership opportunities, you may wish to create the role of Project Coordinator for one student or several students. The Project Coordinator can volunteer, can be appointed by you, can be selected by the class, or can be chosen at random to assist you with the administration and facilitation of the project in the following ways:

- Lead large group sessions
- Serve as liaison between the Action Groups and you or the class and the school administration
- Help to monitor and keep Action Groups on schedule
- Oversee the preparation and distribution of materials, including activity sheets

Community Involvement

Community Participation The Environmental ACTION modules enable students to participate in community service activities and help students develop a commitment to public service. For optimum success, you and your students will want to cultivate the goodwill of all potential participants from the beginning of the project. These participants may include other students, teachers, administrators, librarians, kitchen staff, custodians, maintenance staff, and purchasing personnel, as well as members of the community. The following are some hints that may be helpful in gaining the participation and cooperation of members of your community:

- Notify key support groups in your school and community of your project.
- Provide all participants with a description of the activities and a description of things students will be doing in the course of their work.
- Make a formal request for any specific considerations or resources needed from these individuals.
- Inform students' parents. They may be able to contribute additional resources and/or expertise. Also, parents' understanding of homework assignments such as the home audits will help ensure a positive and productive experience.
- Recognize the cultural, economic, and ethnic diversity of your school community and be sensitive to the varied perceptions of issues related to human health and the health of the environment.

Community Recognition It is important for the students to realize that their responsible investigations of an environmental issue and development of a logical plan for taking action have earned serious consideration by the community. Therefore, it is critical to engage community involvement that will include a thoughtful response to the students' final report and proposal.

One way to encourage community participation is to create an environmental committee at your school. This committee can be set up at the beginning of the project. The members may include parents, teachers, students, administrators, and other school staff. If a school Ecology Club is already in place, you may wish to explore establishing a committee under the auspices of the club. The committee can serve as a resource, a liaison between the students and school officials, and a review body for ideas and suggestions for implementing changes. In order to be effective, the committee will probably need to meet periodically throughout the project to hear status reports and to respond to requests for assistance or advice. You may want to draft a written agreement detailing the role of participants and the service each agrees to provide.

Recognition of students' efforts to bring about positive change in the school setting could include a public award ceremony, certificates of merit, academic or extra-curricular credit, or establishment of an Implementation Committee on which students might serve.

Bringing Role Models into the Classroom

The first activity in each Analyze section suggests that you ask leaders from your community or resource people who are familiar with the school to present information related to the module issue. Plan ahead to find and make arrangements for guest speakers. It can be a time-consuming task, but outside participation in this project is important for a number of reasons:

- The information provided will help students understand how the issue relates to their community, school, family, and themselves.
- Students will be able to ask questions and obtain answers that will assist them in their investigation and audit of the school campus.
- Role models from your community can become valuable and accessible resources for you and your students.

What kinds of role models make effective guest speakers? The main criterion is that they are familiar with the module issue. You may want to invite a school administrator, parent, elected official, resource person with a government agency or advocacy organization, business owner, medical professional, or a family member of one of your students. Your students may have some good ideas about whom they would like to meet and hear speak. You might invite a number of guests to allow students to hear several points of view.

A fun and interactive way to present a speaker is to introduce the guest to your students without revealing the person's connection to the module issue. Then play Twenty Questions, having students ask yes/no questions to try to discover the person's role. Once the role is determined or the maximum number of questions has been asked, you can introduce the guest and provide some background.

Cooperative Learning

Cooperative learning—including small-group discussion, problem-solving, and decision-making activities—is an integral part of the Environmental ACTION curriculum. Because some students may not have previous experience with these processes or may not yet have developed strong skills, you may need to prepare them for this type of interaction. The effectiveness of small group discussion as a learning tool varies widely according to several fairly concrete parameters, and you can influence the outcome in your classroom by sharing these with your students, thus providing a model for them to work toward.

First, we believe that small group discussion is most effective when the number of participants is limited (four to six) to allow members to be aware of and react to one another. Additionally, it is critical that participants have a common understanding of goal(s), have a sense of being part of the group and are willing to work cooperatively, work toward open dialogue and discussion among participants, and recognize that individual accomplishment depends upon the group's success.

Following are some questions and answers about basic cooperative learning terminology and ways to manage small group experiences. Many of these topics are discussed in more detail in *Effective Group Discussion* by John K. Brilhart (Wm. C. Brown Company Publishers, Dubuque, Iowa).

Q: What makes an effective Action Group member?
A: Characteristics include a sense of responsibility for helping the group accomplish its objectives, active participation, willingness to accept tasks, and commitment to the group through good and bad. Also, effective group members offer positive feedback on ideas and information, avoid using words that induce negative feelings, make organized comments, speak concisely, state one point at a time, pose specific questions, and listen courteously and attentively. It is to be expected that students will improve their skills in these areas as they gain experience working in their Action Group.

Roles for individual group members will evolve through the workings of the group. In some cases, there may be no apparent leader but lots of leadership. Conversely, groups may have leaders but lack direction. The best-case scenario is for each group to have a leader who oversees procedures and facilitates the group's process.

Q: How should Action Groups be organized physically?
A: Action Groups work best when they are in comfortable environments that do not offer a lot of distractions. Students should sit close together in a circle and have access to writing surfaces. If you notice some awkwardness when students are first assigned to Action Groups, you might want to introduce an ice-breaker activity to help them get comfortable with working together.

Q: What can I expect from students working in Action Groups?
A: Small groups usually progress through several steps before achieving end goals. Groups vary in terms of the time it takes to move through some or all of the following steps:

- Members of the group work to develop a cohesiveness of goals and understanding of one another.
- The group establishes the identities and roles of individual members, including leadership roles. This step may be accompanied by disagreements.
- The group finds ways to manage disagreements by establishing rules and procedural standards. This usually results in more agreement and an increased feeling of freedom to offer opinions and have them heard.
- The group focuses on the goal(s), and discussion is concentrated on developing solutions to the problems identified.

Q: Is decision-making the same as problem-solving?
A: No. Problem-solving involves a number of phases—identifying the problem; investigating alternatives; developing an implementation plan; and making the necessary changes. Problem-solving sessions should focus on the issue and a careful framing of the problem question before discussing possible solutions. They should also encourage participants to share their knowledge about the issue and refrain from making judgments when discussing alternatives. Decision-making comes into the problem-solving process whenever students choose between various options or possible courses of action.

Q: When is group brainstorming an appropriate process?
A: Brainstorming is an appropriate small group discussion tool when a number of solutions exist for a particular problem. The following conditions of brainstorming should be communicated to participants before brainstorming sessions begin:

- Any idea is valid; the more, the merrier.
- Ideas can build upon previous suggestions.
- Suggestions should not be judged until brainstorming has ended.

Q: What is consensus decision-making?
A: Consensus decision-making is an alternative to the more traditional majority vote group decision-making. In the consensus process, value is placed on each member of the group agreeing to a decision. It may not be the preferred decision of any individual, but it is something everyone can live with. This type of decision-making allows for the group to meet a multitude of criteria for making a choice and for each member to derive satisfaction from the final result. However, consensus decision-making can take considerably longer than other processes such as voting, and in some cases it is not feasible to achieve total agreement.

If you choose to make use of consensus decision-making, you may want to follow certain steps. First, list options, stating reasons in favor of and against them. Next, rank options according to criteria such as accessibility or time constraints. Then, consolidate ideas and summarize the class discussion. Next, narrow down the options to one or two. Using the chalkboard, you might list under each option the pros and cons. Ask for a class vote. If the vote is not unanimous, ask volunteers for each side to explain their reasons. Encourage students to use persuasive arguments but also to listen attentively to the opposing arguments. Continue polling the class until finally all are in agreement about their choice of a study area. Activity sheets are provided for this purpose.

Q: If an Action Group is not staying focused, what can I do?
A: You can help by asking one or more of the following questions:

What are you meeting about? What is your common goal? What do you want to achieve as a group? Are you staying on track with your tasks?

Evaluating Students' Work

There are numerous ways to evaluate students' work and progress. We suggest you begin with the Student Survey provided in the Blackline Masters/Assessment Tools section at the end of this book. This survey contains both objective and subjective questions and can be given at the beginning and the end of the project. If used at the beginning, the process of reviewing the students' answers can serve as a device for discussion or as a tool for assessing your students' base level of knowledge about a particular issue. If used again at the end of the project, the survey can help you evaluate the amount of learning and changed attitudes/behaviors that resulted from students' investigations and activities.

To evaluate different aspects of an individual student's participation in the Action Groups, you may choose to use the Action Group Evaluation Form. This is a simple checklist for your reference. The blackline master is provided in the Assessment Tools section on page 151 of this book. Another evaluation tool that works well in a cooperative setting is a self-evaluation submitted by students. A blackline master Student Self-Evalution Form is provided on page 150 of the Assessment Tools section. In order to best assess long-term performance, you will want to schedule evaluations at regular intervals throughout the project.

What's Next?

Teachers often wonder what happens after their class completes a module and initiates changes or environmental activities at the school. What is next semester's or next year's class supposed to do? As long as the school has areas that need improvement, learning opportunities abound. The following are some options you may find useful:

- Repeat the module activities to evaluate the impact of changes, and continue to develop additional recommendations including modifications of the previous changes.
- Maintain the newly instituted activities at the school.
- Choose another module, and allow your students to explore a different environmental issue.

Environmental ACTION Feedback

We are interested in the ongoing short- and long-term effects of implementing our curriculum at schools, so we invite you to provide feedback about your experience. The data collected will not only provide us with information about the effectiveness of the curriculum, it will allow us to compile information illustrating the degree of change occurring on campuses nationally and abroad. This information also becomes a peer-networking resource for teachers and students who have questions about implementing environmental programs at their schools. We ask that you take a few minutes at the conclusion of this project to complete the Program Evaluation Form located at the end of this guide. We also urge you, at any time, to visit our Web site. The address is: <**http://www.enviroaction.org**>. Let's share our successes and challenges! We'll add your contribution to our site.

About This Module

Teaching Chemicals: Choosing Wisely

Begin by giving students an overview of the module and its activities. Tell them that this project is aimed at reviewing the use of chemicals and chemical products at the school. Explain that they will be studying the costs and benefits of these chemicals on human health and the environment, auditing current chemical use and practices, and looking for alternatives that offer fewer costs or more benefits. Finally, they will be developing a plan for the school that urges adopting some of these "wiser choices" they identified.

Distribute the Student Editions and ask your students to read the Welcome! section. Allow time for discussion of the student books, teacher/student roles, evaluation, schedules, contact with the school administration, procedures, and Student Journals.

To assess students' prior knowledge of this topic before they begin work and to get them thinking critically about chemical products, you may want to have students complete the Student Survey provided in the Blackline Masters/Assessment Tools section of this guide, page 149. (A discussion about using the survey as an assessment tool appears on page 16.)

Since much of the work in this program will be done in small groups, you may want to take time at this point to explain the methodology of Action Groups and offer tips on how to work cooperatively. (See pages 10–12 for suggestions on facilitating Action Groups.)

Time Requirements Most activities are designed to be conducted in a single 50- or 55-minute class session. The second Explore the Issues activity, Define Chemical Products Hazards, may take two sessions. In addition, the first and second Analyze activities may take more than one session, depending on how many guests are invited to speak and how the tours are organized. Also, Action Group work and homework in several of the activities require extra time outside of class. The amount of time assigned to homework and independent group work can vary.

Materials and Preparation Implementation of the *Chemicals: Choosing Wisely* module requires no special materials or equipment. Students will need pencils and paper to take notes and make Journal entries while surveying the school. Poster board, oak tag, construction paper, markers, and general art supplies will help students make effective presentations. In addition, you may wish to obtain chemical product labels (either real or facsimile), Material Safety Data Sheets, and other information about chemical products hazards. (See the Resources section of this book for sources for such information.)

Blackline masters for the activity sheets are located at the end of this book. You may choose to photocopy all of them before beginning a new module, or one at a time at the appropriate point in the lesson. See the Prepare section of each activity for any additional preparation required.

Explore the Issues Activities

The Explore activities emphasize "the big picture," presenting data for students to analyze and use to create graphs, and offering ideas and information about synthetic chemicals. Students become aware of both the benefits and hazards of chemical products, and begin to look at some alternatives that have proven effective and safer. The activities give students experience working independently, in small groups, and as a class.

Activity 1: Evaluate the Growth and Uses of Chemical Products Students read about the growth and uses of chemical products since World War II. They analyze data in a table showing the growth and then use the data to make a graph.

Activity 2: Define Chemical Product Hazards Students study a chart showing some of the ways chemical products can be hazardous. Then they identify some possible chemical hazards in a variety of situations.

Activity 3: Understand Chemical Product Labeling Students learn about the various kinds of information on chemical product labels and apply that knowledge in analyzing and comparing the relative safety of cleaning products.

Activity 4: Investigate Alternatives Students learn about some of the health and environmental hazards of pesticides, insecticides, and herbicides and then test their own tolerance for pests. Then they look at some healthier choices offered by an alternative approach to pest control called Integrated Pest Management (IPM).

ACT Activities

Once students have an overview of environmental and health issues related to the use of synthetic chemicals, they embark on the ACT activities. These activities guide students through the process of conducting a chemical-use audit of the school (Analyze activities); investigating alternatives to the use of chemicals (Consider Options activities); and proposing changes (Take Action activities).

Analyze Activities

Activity 5: Learn About Chemical Products in Your Community Students listen to information presented by one or more experts from the community which focuses on issues in three categories of chemicals: pesticides, insecticides, and herbicides; cleaning products; and building maintenance materials.

Activity 6: Tour Your School Students gather and record information from presentations and tours by school staff members with expertise and/or responsibility for using chemical products at the school.

Activity 7: Plan Your Chemical Use Audit Students plan their audit to identify the primary uses of the three categories of chemicals at the school, organizing the work by type of chemical or by site. Students are divided into Action Groups and begin work on their Action Group Plans.

Activity 8: Conduct Your Chemical Use Audit Members of Action Groups, using observation, interviews, and other sources to collect the information necessary to complete their Chemical Use Audit Forms.

Analyze 9: Summarize Findings Students meet in their Action Groups to combine, organize, and summarize the information they have collected. Groups share their findings with the class.

Consider Options Activities

Activity 10: Brainstorm Alternatives Action Groups explore ideas they have learned about in the course of their work and research to find additional alternatives to the chemical products they have targeted. These alternatives can include changes in products, changes in the physical environment, or changes in attitudes or behaviors that will allow for healthier choices.

Activity 11: Weigh Costs and Benefits of Alternatives Students evaluate the alternatives to using harmful chemical products that they generated in Activity 10, weighing costs and benefits. Both monetary and nonmonetary costs are considered, as well as short-term and long-term benefits.

Activity 12: Make Recommendations Action Groups review their suggestions for alternatives and select those they wish to recommend to the class. They then develop a presentation that includes persuasive arguments and supporting data represented by charts, graphs, and other information.

Take Action Activities

Activity 13: Choose Alternatives Representatives from Action Groups present their recommendations for alternatives to current chemical use and practices at the school. Students evaluate the presentations, and then the class decides which to include in the final proposal.

Activity 14: Prepare Proposal Students meet together as a class to outline their final proposal. The process includes completing a checklist to identify who does what part of the proposal—introduction, description of recommendations, charts, graphs, conclusion, and so on. Students then take responsibility for completing their part of the proposal, and it is submitted to the school administration or appropriate group within the school.

Activity 15: Follow Through In this last activity of the module, students assess the response to their proposal. They check on its implementation and plan how to maximize the impact of their recommendations. Ideas for continuing and expanding the alternative measures are addressed.

Assessment Tools

The following assessment tools have been provided in the Blackline Masters/Assessment Tools section of this guide: Content Quiz, Student Survey, Student Self-Evaluation Form, and Action Group Evaluation Form. At this time, you may wish to administer the Content Quiz and the Student Survey as pre-tests for the module. These two tools are intended to be used again, as post-tests, at the end of the module. A reminder has been included on page 81 of this guide. Answers to the Content Quiz are shown below. The Student Survey (see page 149) is composed of subjective questions, so answers will vary. The Student Self-Evaluation Form and the Action Group Evaluation Form can also be administered at the end of the module. Finally, an optional Program Evaluation Form has been provided for your feedback.

Answers to Content Quiz

1. b. false. Scientists create 500 to 1000 new synthetic chemicals each year.
2. a. true
3. c. both of the above
4. a. true
5. c. easily catch on fire
6. b. There may still be some residuals in the container.
7. b. false
8. b. false. By law only the "active" ingredients must be listed (for example, in bug sprays, the active ingredient is the one that kills the bugs). Most chemicals are composed of many chemicals in addition to the active ingredients—and many of these unlisted ingredients, called "inert" ingredients, are hazardous and represent threats to human health and the environment.
9. b. false. Unfortunately, information about how to classify chemicals as being hazardous or not is confusing and incomplete. The U.S. government has only begun to require information about the hazards of chemicals in recent years. But the majority of chemicals have not yet been completely tested.
10. b. false. Federal law only requires products that are known to be hazardous to be labeled as such. Many companies do warn of the effects if swallowed but do not mention problems caused by inhalation or skin contact.
11. c. weeds. Insecticides are used for insects, herbicides for plants, fungicides for fungi. All of the above are considered pesticides.
12. a. true

Cross-Curricular Suggestions

Description of Activity	Curriculum Connections	When to Use
Write the following headings on the chalkboard: 1500, 1800, 1850, 1900, 1925, 1950, and Today. Encourage students to find out about the kinds of products people used to clean and maintain their homes and manage pests and insects. When students have collected as much information as possible, you might have them share information and make a timeline for each category of use showing when new products or technologies were introduced. For example, in 1500, dogs and cats may have been the most common form of rodent control. When was the mousetrap invented? Poison? What kind of control is used today? Did people paint their houses in 1500? When was whitewash first introduced? What was it made of? Why was it used? Use the discussion to raise issues about how changing tolerances and the awareness of health risks or perceived health risks have contributed to the adoption of new products and technologies.	History/Social Studies, Science (Technology), Language Arts	Following Activity 1

Description of Activity	Curriculum Connections	When to Use
Have students get statistics on chemical usage in at least ten other countries. They can then convert the data into graphs showing use relative to population and use the graphs as a basis for discussion. Issues such are the following might be addressed: • How does the rate of use compare between industrialized nations such as the United States and developing nations? • Why do other industrialized nations show rates of use that are markedly different from that in the United States?	Math, History/Social Studies, Language Arts	Following Activity 4
Ask each Action Group to draft, revise, and send a letter to one of the following agencies: a toxic waste company, an environmental group active in the area of chemical hazards, a local water quality group, or a local utility in charge of disposal of hazardous materials used by customers. Students' letters should request information about chemical hazards, safe use and disposal, alternatives, and relevant laws and regulations. Replies can be shared and the information used to help develop alternatives and make cogent arguments to support the recommendations they will make to the school later.	Language Arts	Following Activity 7

Description of Activity	Curriculum Connections	When to Use
Discuss with students how the combination of some common chemicals—for example, ammonia and chlorine—can result in highly toxic substances. Have students find out the chemical reaction, the resulting substance, and the hazards. Encourage students to identify how these two cleaning agents might come in contact with one another in their homes, and what precautions should be taken to avoid that happening.	Science, Language Arts	Following Activity 9
Write the following limerick on the chalkboard: I stepped in some chemical goo, And got a big glob on my shoe. It ate through the sole And it made a big hole In my sock, and it burned my foot, too! Read the limerick aloud and discuss the types of meter and rhyme scheme which characterize the limerick form. Then encourage students to write environmental/chemical limericks of their own using this model. You might arrange to have them posted around the school to raise awareness about what the class has been doing among other students and staff and to lay some groundwork for gaining support for the changes that your students will be recommending.	Language Arts	Following Activity 12

Description of Activity	Curriculum Connections	When to Use
Ask students to bring in articles from newspapers or magazines reporting on issues related to chemical hazards. Invite them to put together a portfolio of the articles, adding to it whenever new information appears. The portfolio can be used as background to help bolster their recommendations to the school regarding changes in chemical use.	Science, Language Arts	Following Activity 14

Explore the Issues

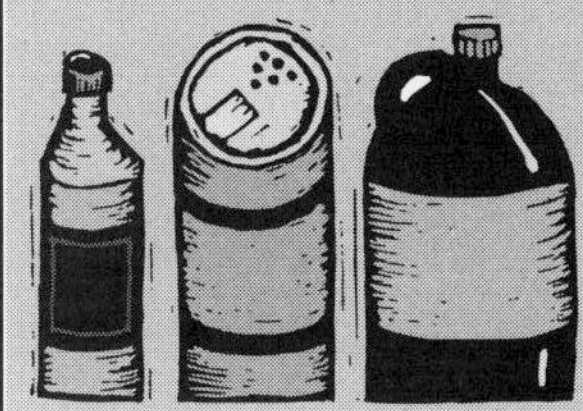

EVALUATE THE GROWTH AND USES OF CHEMICAL PRODUCTS

EXPLORE 1

Students analyze information about the growth and uses of synthetic chemicals since World War II. They analyze data in a chart and use it to create a line graph. They then make predictions about future production of synthetic chemicals.

Key Concepts

- The use of synthetic chemicals has increased rapidly since World War II.
- Synthetic chemicals are used in many everyday products.
- Synthetic chemicals have benefits.
- Synthetic chemicals can get into our air, soil, and water.

Objectives

After completing this activity, students will be able to

- describe the growth of synthetic chemicals since 1950
- name some everyday products that have synthetic chemicals in them
- list some benefits of synthetic chemicals
- describe some of the ways synthetic chemicals can enter the environment

Prepare

Depending on your students' awareness of synthetic chemicals, you might want to gather some synthetic chemical products to display in your classroom. You might display some natural chemical products as well, such as a thermometer (mercury), baking soda, and so on. Invite students to browse and look at the ingredients lists before you move into Setting the Stage.

Materials

Activity Sheet 1 for each student (See Blackline Masters section.)

Setting the Stage

Discuss briefly the questions on student page 9.

- **What are chemicals?**

  Answers will reflect students' prior knowledge.

- **What are chemicals used for?**

  Responses will vary based on students' prior knowledge.

Vocabulary

chemical
herbicide
insecticide
pesticide
petrochemical
synthetic chemical

Focus

1. Continue the class discussion about chemicals and their uses by having students read the material on synthetic chemicals and answer the questions on the student page.

- **About how many kinds of synthetic chemicals are in use today?**

  70,000

- **What are some everyday products that contain synthetic chemicals?**

  Examples include pharmaceuticals, pesticides, herbicides, insecticides, floor wax, wax remover, cleaning products, building materials, carpeting, wallpaper, personal care products.

- **What are some of the benefits of synthetic chemicals?**

  They can fight disease; kill rodents, bugs, weeds, bacteria, and mold; make things waterproof and fireproof; and so on.

- **What are some of the ways that chemicals get into the environment?**

  They can be released into the air or find their way into the water supply and soil.

2. Guide students to greater understanding about the proliferation of synthetic chemicals and their use by asking questions that invite them to draw conclusions from what they read. For example, you might ask the following:

- **Why do you think synthetic chemicals have been in such great demand since World War II?**

  People like the convenience and efficiency that they seem to provide.

- **If scientists keep creating new synthetic chemicals at the current rate, what will happen?**

  Many more chemicals will be absorbed into the air, water, and soil cycles and their resulting impact on human health and the environment will increase.

3. Explain to the class that they will learn more about synthetic chemicals and how the production of them has increased since 1945. Have students work in pairs or small groups to complete Activity Sheet 1.

It's a Wrap

Review Activity Sheet 1. You may wish to elaborate on the data provided in the chart. Inform students of these facts: Synthetic chemicals are increasing not only in number of tons produced but in number of different chemicals produced. Scientists add 500 to 1000 new synthetic chemicals each year.

Invite students to share their thoughts and concerns regarding the benefits of synthetic chemicals and the questions they have concerning their use.

Home

Assignment Home activity on student page 10.

Interview someone in your family or neighborhood who remembers growing up before the 1950s. Ask them what they remember about pesticides, cleaning products, and other materials that are now made with synthetic chemicals. How many did they use in their home? What did they use instead of the products that are used today? In your Journal, write about what you learned from your interview.

ACTIVITY SHEET

EXPLORE 1

Name

THE GROWTH OF SYNTHETIC CHEMICALS (part 1)

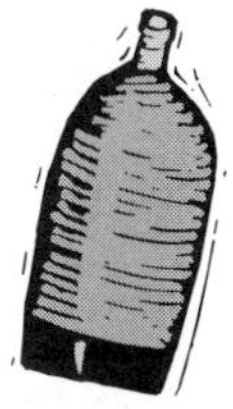

Make a line graph based on the information presented in the chart below. Then answer the questions.

Approximate Amount of Synthetic Chemicals Produced in the United States (in Tons)

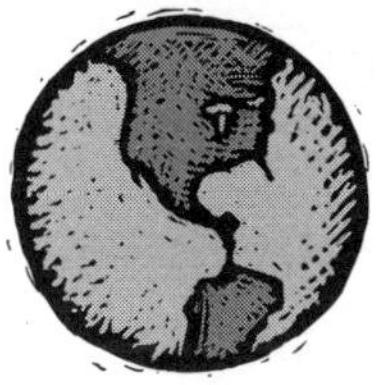

1945	8 million	1970	47 million
1950	8 million	1975	70 million
1955	12 million	1980	100 million
1960	18 million	1985	102 million
1965	45 million		

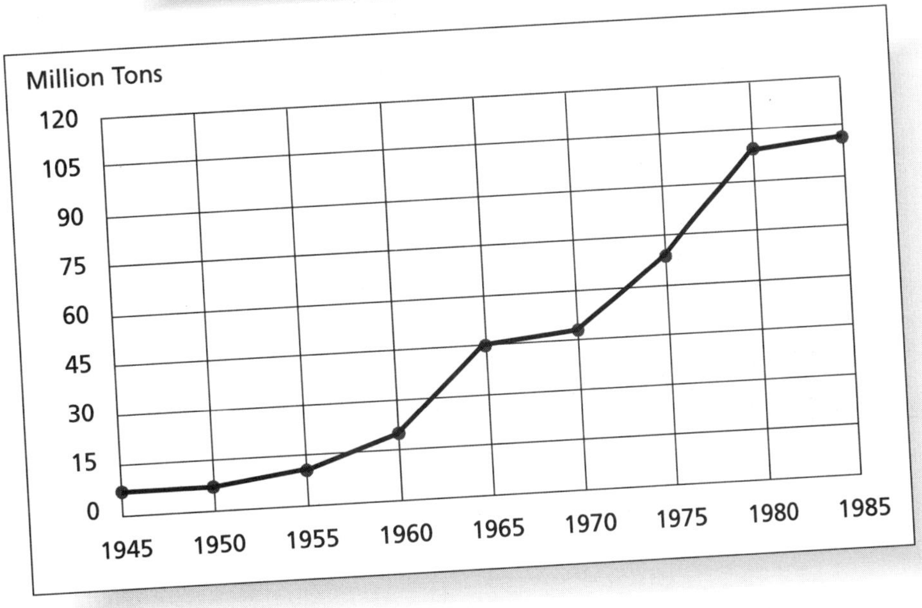

1. For every ton of synthetic chemicals produced in 1945, about how much more was produced in 1980? Circle the correct answer.

Twice as much Five times as much Twelve times as much

ACTIVITY SHEET

EXPLORE

Name

THE GROWTH OF SYNTHETIC CHEMICALS (part 2)

2. From 1950 to 1980, which decade had the smallest increase in production of synthetic chemicals?

1950 to 1960

3. The line graph you drew stops at 1985. Look at what happened between 1980 and 1985. If that same trend were to continue, about how many million tons of synthetic chemicals would be produced in 2010?

About 112 million tons

Explain your reasoning.

An increase of 2 millon tons every 5 years for 25 years means an increase of 10 million tons over the 102 million tons in 1985.

DEFINE CHEMICAL PRODUCT HAZARDS

Students consider some of the hazards associated with chemicals. They study a chart and answer questions about ways in which products can be hazardous to health and to the environment and then identify and explain possible chemical hazards shown in an illustration.

Key Concepts

- Some chemicals present health and environmental hazards.
- Chemicals that are toxic can cause specific, serious, and long-lasting problems.
- Chemical products are considered to be hazardous if they are flammable, explosive, corrosive, and/or toxic.

Objectives

After completing this activity, students will be able to

- state the ways in which chemicals can be hazardous
- recognize potentially hazardous uses of chemicals

Prepare

You might want to read the background material on the health and environmental risks of chemicals, which you can find in Issues and Information sections A, D, E, and F in the back of the book.

Materials

Activity Sheet 2 for each student

Setting the Stage

Discuss briefly the questions on student page 12.

- **What precautions might you want to take when handling straight lemon juice?**

  Students will recognize that it would burn if they got the juice on a cut or in their eye and may add that it is the acid (a chemical) in the lemon that causes this reaction.

- **What would you have if you diluted the lemon juice and added sugar?**

  Lemon juice, water, and sugar make lemonade. Help students recognize that everything is made up of chemicals; some can cause harm if they are not used appropriately or carefully.

- **What are some of the possible hazards of chemical products?**

  Students will share what they know about possible health and environmental hazards of chemical products. Generalize the discussion to make the point that chemical products need to be used, handled, and disposed of appropriately.

Vocabulary

genetic mutation
hazardous
reactive
toxic

Focus

1. Continue the class discussion by having students study the chart and answer the questions on the student page.

- **The oven cleaner that Mona uses is toxic. What precautions do you think she should take when she uses it?**

  Possible responses: she should avoid inhaling fumes and getting oven cleaner in her mouth;

she should wear rubber gloves and goggles; she should keep windows open to allow cross-ventilation.

- **Byron stores ant and roach killer, oven cleaner, and furniture polish in a cabinet over his stove because he wants to keep them out of the reach of children. All of these products are flammable. What advice would you give to Byron?**

  Possible response: he should find a different storage area that is out of the reach of children, but it should not be near anything that might cause the products to get too hot or be exposed to open flames.

- **Samantha puts an almost-empty spray can of insecticide into her trash. The product is explosive under certain conditions. Why is Samantha's action dangerous to herself, her neighbors, and the trash collectors?**

  Possible responses: If too much trash is piled on top of the can, it might cause too much pressure, and the can might explode. If the trash sits out in the sun, it might get too hot, and the can might explode. When the trash collectors pick it up, the can might be punctured in the compactor; this could also cause an explosion.

- **While his dog watches, Marty applies paint remover to some old metal patio furniture. The phone rings before he can wipe it off, and he forgets about it for a few hours. The product is corrosive. What do you think Marty will find when he goes back out to the patio?**

  Possible responses: The patio furniture will probably be corroded; the dog might be injured.

2. Explain to the class that they will learn more about how chemicals can be hazardous. Have students work in pairs or small groups to complete Activity Sheet 2.

It's a Wrap

Review Activity Sheet 2. You may wish to have students compare their answers. Invite them to add their own ideas about chemical hazards to the illustration, based on observations they have made in their own surroundings.

Discuss students' answers to It's a Wrap questions.

- **What precautions should be taken with products that are flammable?**

  When they are being stored, they should be kept out of the reach of children, in a cool, dry place away from open flames and other sources of heat. When they are in use, care must be taken to keep them away from fire or too much heat.

- **Name three things that can cause an explosive product to detonate.**

  Any three of these: heat, sudden shock, pressure, incompatible substances.

- **What are two ways that toxic products that we use today can be harmful to people who haven't even been born yet?**

  They can cause genetic mutations and fetal harm. They can also affect the environment, contaminating air, soil, and water for generations to come.

- **Name some everyday products used in your home that can be hazardous.**

  Possible answers: rat poison, ant killer, flea spray, flea collars, snail pellets, oven cleaners, floor wax, furniture polish, paint, paint thinner, paint remover, chlorine bleach, some art supplies.

Home

Follow-Up Invite students to share what they learned during their interviews and to make some generalizations based on the combined information.

Assignment Home activity on student page 14.

Tour your home and list in your Journal all the places where you find a pesticide, insecticide, herbicide, or a cleaning or maintenance product.

ACTIVITY SHEET

EXPLORE

Name

HAZARDS OF CHEMICAL PRODUCTS (part 1)

Study the picture and find as many possible sources of chemical pollution or hazards as you can. Circle each one, and put a number inside each circle. Then use the lines to list the numbers and explain how each circled hazard might affect humans, animals, and/or the environment.

Students may have found the following sources of possible hazards: crop-dusting plane (children in bus and animals in the field can inhale the pesticides, pesticides can contaminate food crops, pesticides can run off into the river); factories (may be spewing chemicals into the air and dumping them into the river); pesticides on the roses (may drift into the neighboring yard and harm the baby or pet); varnish and thinner just thrown into the trash pose a potential hazard to trash collectors and the environment; (discarded oil can get into soil and water, threatening fish, birds, and other wildlife that use the water).

ACTIVITY SHEET

EXPLORE

Name

HAZARDS OF CHEMICAL PRODUCTS (part 2)

UNDERSTAND CHEMICAL PRODUCT LABELING

EXPLORE 3

Students learn about the requirements for and limitations of product labels. They interpret sample product labels and recommend alternatives to products that are hazardous.

Key Concepts

- Product labels are a source of certain kinds of information about hazardous chemicals.
- Product labels use the words *Poison, Danger, Warning,* and *Caution,* each of which indicates a different degree of hazard.
- Alternatives to using hazardous products are available.

Objectives

After completing this activity, students will be able to

- interpret warnings on product labels, including the fact that not all hazards are identified
- suggest alternatives for hazardous products

Prepare

Collect labels or label information from a variety of products for students to use in the Focus section below. Read background information about hazardous materials classification and about alternative cleaning products in Issues and Information sections A and B.

Materials

Activity Sheet 3 for each student

Setting the Stage

Discuss briefly the questions on student page 16.

- **What kinds of warnings do you recall seeing on product labels?**

  As students answer, record their responses on the chalkboard. For example: Keep out of reach of children. Use with adequate ventilation. Harmful or fatal if swallowed. Store in a cool, dry place.

- **Do you think the direction "Use with adequate ventilation" is clear? Explain your answer.**

  Responses should focus on the vague and subjective nature of the phrase "adequate ventilation."

Vocabulary

active ingredients
carcinogen
corrosiveness
flammability
inert ingredients
surfactant
toxicity

Focus

1. Continue the class discussion by having students read the material and answer the questions on the student page.

- **Which is more hazardous—a product labeled "Danger" or a product labeled "Caution"?**

  A product labeled "Danger" is more hazardous.

- **Why do you think the law requires a list of the common or chemical names of the hazardous active ingredients?**

  In case of a medical emergency, a physician needs to know what chemical may be causing the problem in the patient.

- **Why do you think labels don't have to list possible carcinogens?**

  Unless a substance is listed as a carcinogen by the Environmental Protection Agency, which regulates many pesticides and cleaning and maintenance products, companies are not required to list the substance as a possible hazard. Whether or not a substance is carcinogenic is difficult to establish, and testing may be inconclusive or incomplete.

- **What improvements to product labels would you recommend?**

  Clearer directions about proper and safe use, safe disposal of unused portions, and a listing of inert as well as active ingredients. Manufacturers are not currently required to specify ingredients that may reveal trade secrets other than to identify them as being active or inert.

Guide the class to greater understanding about product labels by displaying in the classroom the labels you collected. Discuss them and invite students to compare them to see which give the clearest directions for use. Ask students to consider questions such as the following:

- **If you had a choice between a product that had no warnings on the label and one that had 350 words of warning, which one would you choose?**
- **If a product label says that the product is hazardous to humans and domestic animals, what do you think the product might do to air, soil, and water if it is used or disposed of incorrectly?**

2. Explain to the class that they will learn more about the chemicals listed on product labels. Have students work in pairs or small groups to complete Activity Sheet 3.

It's a Wrap

Review Activity Sheet 3.

Discuss students' answers to It's a Wrap questions.

- **What four words are used on product labels to indicate degrees of hazard? Which one is used to indicate the most serious, immediate hazard?**

  The four words are *Poison, Danger, Warning,* and *Caution.* The one that is used for the most serious level of hazard is *Poison.*

- **What are some of the concerns you have about warning labels?**

  Responses may include the failure to list the hazards of inert ingredients, the lack of testing of ingredients for toxicity, the lack of clarity about safe use and disposal, and so on.

- **Why do you think people continue to use products that have hazardous chemicals in them?**

  They might not be aware of the dangers; they might not bother to read the labels; they might think these products do a better job than the safer and/or cheaper alternatives; they might be influenced by advertising.

Home

Follow-Up Discuss students' lists of chemical products and hazards. Encourage them to explore whether their attitudes about using these products changed after learning about some of the hazards they pose.

Assignment Home activity on student page 18.

With a parent or other family member, go to a store that sells cleaning supplies. Together, read the labels on at least five products designed for the same purpose (such as laundry detergents, stain removers, dish detergents, or window cleaners). Try to find one product that is not hazardous to humans or to the environment. Make a list of all the products you looked at, rank them in order from least hazardous to most hazardous, and circle the one you think your family should use.

ACTIVITY SHEET

EXPLORE 3

Name ____________________

READING PRODUCT LABELS (part 1)

The warnings below are from actual product labels. The first label is for a product that removes mildew from certain surfaces. The second label is for a dish-washing detergent. Study the labels and then answer the questions.

PRECAUTIONARY STATEMENTS: Hazardous to humans and domestic animals.

WARNING: CAUSES EYE AND SKIN IRRITATION. Do not get in eyes or on clothing. Vapors may irritate. Use only in well-ventilated areas. Open windows and turn on fans before use. If the vapors bother you, leave the room while the product is working. Avoid prolonged breathing of vapors. Not recommended for use by persons with heart conditions or chronic respiratory problems such as asthma, emphysema, or obstructive lung disease. Harmful if swallowed. For sensitive skin or prolonged use, wear gloves. **STATEMENT OF PRACTICAL TREATMENT: EYE CONTACT:** Remove contact lenses and rinse eyes thoroughly with water for 15 minutes. **IF SWALLOWED:** Drink a glass of water and call physician. **IF BREATHING IS AFFECTED:** Get fresh air immediately. **SKIN CONTACT:** Remove contaminated clothing and wash affected area. **PHYSICAL AND CHEMICAL HAZARDS:** This product contains bleach. Do not use or mix with other household chemicals such as ammonia, toilet bowl cleaners, rust removers, vinegar, or acid. To do so will release hazardous gases. **STORAGE AND DISPOSAL:** Store in a cool place. Do not reuse empty container. Wrap in a plastic bag and discard.

CAUTION: May cause burns to skin, eyes, and mucous membranes. In case of external contact, flush with water for 15 minutes and get immediate medical attention. If swallowed, give a glassful of water or milk and call a physician. **KEEP OUT OF REACH OF CHILDREN.**

1. Which label indicates a more hazardous product?

   The first one, for the mildew remover, because its "Warning" designation indicates a more serious level than "Caution."

ACTIVITY SHEET

EXPLORE

Name

READING PRODUCT LABELS (part 2)

2. For the first product, the mildew remover, what problems might occur in a small, windowless room?

In a small, windowless room, it would be difficult to provide "adequate ventilation." An open door and a fan might not provide enough cross-ventilation. The user of the product could easily inhale too much of the vapors. The biggest problem is that mildew often builds up in bathrooms, especially in bathrooms that don't have windows. The very room where one would be most likely to want to use a mildew remover is the type of room where it is least safe to do so.

3. For the second product, the dish-washing detergent, what should you do if you get some in your eyes?

Flush with water for 15 minutes and get immediate medical attention.

4. What alternative products could you substitute for the mildew remover?

Ammonia, baking soda, borax, or washing soda

5. What alternative products could you substitute for the dishwashing detergent?

Washing soda

INVESTIGATE ALTERNATIVES

4 EXPLORE

Students consider some of the hazards of pesticides, insecticides, and herbicides. They contribute to a survey about their own level of tolerance for ants. They interpret the data contributed by the class and organize the information in a chart. Finally, they identify alternatives to chemical pest control.

Key Concepts

- Pesticides, insecticides, and herbicides are hazardous not only to the pests they are meant to control, but to people, pets, wildlife, and the environment.
- Overuse of pesticides, insecticides, and herbicides has produced pests that are resistant to the chemicals.
- Alternatives to pesticides, insecticides, and herbicides are available.

Objectives

After completing this activity, students will be able to

- recognize the downside of using pesticides, insecticides, and herbicides
- identify alternative methods for controlling pests, insects, and weeds

Prepare

Review the material about alternatives to pesticides, insecticides, and herbicides in Issues and Information section C, "Integrated Pest Management."

Materials

Activity Sheet 4 for each student

Setting the Stage

Discuss briefly the questions on student page 20.

- **What are some pests that are commonly seen in gardens and around houses, apartments, and schools?**
- **What are some of the usual ways of dealing with these pests?**

As students answer the questions, write their answers on the board in two columns. Possible responses:

Common Pests	Ways of Dealing with Them
ants	ant poison; ant spray; sprinkling mint, cayenne, red pepper, borax, or boric acid along ant trails; squeezing lemon juice or soapy water at place of entry; caulking cracks in buildings
mice	poison; mouse traps; cats; removing any food they might be attracted to; cleaning up debris
cockroaches	poison; caulking small cracks around shelves, cupboards, baseboards, pipes, sinks, and bathroom fixtures; sprinkling boric acid in places where cockroaches hide; cleaning up debris

Vocabulary

integrated pest management
organic
organically grown

Focus

1. Have students read and discuss the paragraph. Then have them do the activity, "Testing Your Pest Tolerance." Ask them to write their responses to complete Activity Sheet 4.

2. Help students complete Activity Sheet 4 by reading aloud each statement regarding pest tolerance. Have students who responded to the statement raise their hands when it is read. Students tally and total the number of hands raised and record the information on their charts. Have students work in pairs or small groups to answer the questions on the second part of Activity Sheet 4.

It's a Wrap

Review Activity Sheet 4. Discuss the class's general tolerance level for ants. You may wish to expand the discussion by talking about their tolerance level for other pests, such as mice, flies, fleas, and cockroaches. Ask them how their tolerance level changes for pests that might carry diseases. Encourage students to recognize the benefits of increasing their tolerance for certain kinds of pests.

Discuss students' answers to It's a Wrap questions.

- **Why are chemical pesticides, insecticides, and herbicides not really effective in the long run?**

  The pests that do manage to survive the chemical treatment produce offspring that are resistant to the chemical; eventually, the chemical is no longer effective against that particular pest, and stronger chemicals need to be used; those stronger chemicals in turn produce even more resistant pests.

- **What are some of the dangers of using chemical pesticides, insecticides, and herbicides?**

  The chemicals get into the air, soil, and water, affecting people, wildlife, and the environment in ways that were not intended. Some chemicals that are nontoxic on their own can become hazardous when mixed with other chemicals. Overuse of pesticides can have cumulative or long-term effects.

- **What would you say to someone who didn't want to eat an organically grown (without the use of pesticides) tomato because it didn't look perfect?**

  Possible response: You don't know what chemicals are stored in a tomato that has been treated with pesticides. It might look perfect but be hazardous to your health, while an organic tomato might look damaged but be healthier for you.

- **List all the benefits and drawbacks you can think of for tolerating pests such as ants. What can you conclude?**

  Examples of benefits might include reduced chemical exposure and a corresponding reduction in health problems related to exposure; fewer chemicals in our air, water, and soil; fewer chemical-resistant insects. Examples of costs might include the presence of more ants, the possible inconvenience of having to be more careful about food storage and clean-up, and the potential damage to wood buildings from specific kinds of ants (carpenter ants). Students may conclude that the benefits outweigh the costs.

Home

Follow-Up Invite volunteers to share what they found out when reading the labels of cleaning products. What were the most hazardous products and product categories? What were some products that were healthier choices? Encourage students to discuss the changes they would like their families to make.

Assignment Home activity on student page 22.

Ask your family members about their tolerance for pests. You might want to conduct the same activity with your family as you did with your classmates. Record their answers in your Journal.

ACTIVITY SHEET

4 EXPLORE

Name

RECORDING DATA

Use the chart below to record data from the pest tolerance test. Read the statements in column 1. In column 2, make a mark for each person in the class who responds to that statement. Count the marks and write the total in column 3. Then answer the questions.

Pest Tolerance Test

Statement	Tally	Total
I would use a chemical pesticide if I found . . . • ants on the windowsill	**Tallies and totals will vary.**	
• ants crawling out of cracks in the wall		
• ants crawling on the floor		
• ants on my desk		
• ants inside my backpack or book bag		
• an ant crawling up my leg		
• an ant in my hair		
• ants crawling on the food in my lunch bag		
I would not use a chemical pesticide in any of these situations		

Without using pesticides, how can you . . .

• keep ants from crawling on a windowsill? **Possible responses: locate the place of entry and squeeze a lemon onto it, leaving the peel; sprinkle cayenne, red pepper, or boric acid along ant trails; put coffee grounds or crushed mint leaves around the windows or other points of entry.**

• keep ants from crawling out of cracks in a wall? **Fill cracks around shelves, cupboards, sinks, and bathtubs with silicone caulking, putty, or paint.**

• keep ants out of the food you bring for lunch? **Store all food in containers with tight lids; clean the inside and outside of your lunch box after each use.**

Analyze

LEARN ABOUT CHEMICAL PRODUCTS IN YOUR COMMUNITY

By listening to a presentation by a community resource person or persons, students get an overview of chemical use, hazards, disposal, laws, and product trends. Students may also learn about some specific materials or products that can be substituted for more toxic ones, processes and procedures that can reduce the need for hazardous products, or ways to reduce the effects of hazardous materials on humans and the environment.

Objectives

After completing this activity, students will be able to

- identify some chemical issues affecting their community
- list some specific chemicals or types of chemicals that are of concern
- describe some of the health and environmental risks certain chemicals can pose

Prepare

Arrange for a representative or representatives from a local business, government agency, hazardous materials handling/disposal entity (public or private), or environmental group to talk to your class about chemical issues, highlighting local concerns. Make sure that the person is prepared to discuss a variety of chemicals, their hazards, their short- and long-term effects, and some safer/healthier alternatives (where available).

Please note that while students will not be formally investigating alternatives until the next section, Consider Options, it will be useful for them to learn about and record any alternatives that are presented during the course of the Analyze section. This lesson represents the first such opportunity to learn about alternatives.

You might want to provide a copy of the annotated Activity Sheet 5 ahead of time for the speaker to use as a guideline and briefly describe the Explore activities students will have completed prior to the guest's appearance so that he or she can get a sense of the background students already have. Encourage the speaker to allow time for questions from students.

Explain that the purpose of this presentation is to make students more aware of the many chemical products they come in contact with that are harmful in one way or another and suggest some alternatives to these materials. Encourage the speaker to focus on things that students might be most likely to come in contact with in the school environment. It is important that this presentation not overwhelm students with the dangers of hazardous materials, but rather focus on a few key categories of hazards and positive things students can do to make a difference.

Note that while student activities in this module are focused on three categories of chemical products—pesticides, insecticides, and herbicides; cleaning materials; and maintenance products, such as paints and solvents, varnishes, waxes, and polishes—you may feel that your class is able to take on a wider range of materials or that there is significant interest in additional categories or that there are chemical products of particular concern in your community or school. For example, you might choose to have your class explore art supplies; building, decorating, and remodeling materials; personal care products; chemicals used in the science classrooms. If you do decide to expand the focus of the module, the organizations and resources listed in the back of the book can provide useful information.

See the Focus section for information to present to students before the resource person's presentation.

Materials

Activity Sheet 5 for each student (See Blackline Masters section.)

Setting the Stage

Discuss briefly the questions on student page 27.

- **What are some of the chemical issues affecting your community?**

  Answers will reflect local concerns. Possible answers: toxic waste sites; pesticide, herbicide, and insecticide runoff from agricultural lands; heavy metals; toxic by-products of local industries; gas stations; cars and other vehicles.

- **What are some of the chemicals or types of chemicals that are of concern?**

  Possible answers: pesticides, insecticides, and herbicides; heavy metals; cleaning and maintenance products.

- **What are some of the environmental and health risks that hazardous chemicals may pose?**

  Risks to humans (depending on the exposure and material) might include burning of skin or eyes, difficulty in breathing, vomiting, nervous disorders, brain damage, damage to other organs, increased numbers of cancers, birth defects and miscarriages; pollution of air, water, and soil (and the accompanying effects on wildlife); cumulative damage to health and the environment.

Point out that much of the information students gain about hazards in the community will be helpful when they investigate the use of chemical products on the school campus later on.

Vocabulary

solvent

Focus

1. Before the community resource person comes to class, explain that students will be learning about chemical issues that are of concern locally. They will learn to identify some hazardous products, their use, and their effects. In addition they may hear about some possible alternatives to these hazardous products. All of this background will prove useful when they tour the school in Activity 6.

2. Distribute Activity Sheet 5 and introduce the guest speaker(s) to the class. Encourage students to ask questions and take notes during the presentation. Suggest that although they may not be able to fill in all the columns for all the products, they should fill in as much as they can. Students can use a separate sheet of paper if they need more space to record information.

It's a Wrap

Review Activity Sheet 5. Have students share their findings and revise their charts. They may add information they gained in the Explore activities to what was presented by the speaker if they choose. Instruct students to save their completed activity sheets to refer to as they plan and conduct their school audit.

If students do not have time to write paragraphs, have volunteers share what they learned from the presentation about materials they routinely come in contact with (directly or indirectly) that are or can be hazardous.

Home

Follow-Up Have students share what they discovered about their families' tolerance for pests. If you wish, you can have the students tally the family responses in the same way they tallied their own responses.

Assignment: Home activity on student page 28.

With a parent or another adult at home, make a list in your Journal of all the products your family uses that fall under these categories: pesticides/insecticides/herbicides, cleaning materials, and maintenance products (paints, varnishes, solvents, waxes and polishes). Next to each item on your list, write one condition under which the product could be hazardous.

ACTIVITY SHEET 5 ANALYZE

Name

CHEMICAL PRODUCTS IN THE COMMUNITY (part 1)

Use this sheet to record information as you listen to the presentation by the resource person from your community.

Pesticides, Insecticides, and Herbicides

Product/ Material	Use	Health and Environmental Risks	Alternatives (if presented)
herbicide	kill weeds	air, water, and soil pollution; skin rash, flu-like symptoms, headaches, difficulty breathing/cancer, birth defects	hand weeding, boiling water, planting alternatives to lawn, using physical barriers, using less toxic products
pesticide	kill mice	same as above	sealing entry holes, removing garbage that attract pests, using less toxic products
insecticide	kill ants	same as above	increased tolerance; see above

ACTIVITY SHEET 5 ANALYZE

Name ____________________

CHEMICAL PRODUCTS IN THE COMMUNITY (part 2)

Cleaning

Product/ Material	Use	Health and Environmental Risks	Alternatives
ammonia	clean, disinfect	skin rashes, difficulty in breathing, toxic gas (from mix of ammonia and chlorine), water pollution	vinegar, baking soda, hot, soapy water, mixture of water and non chlorine bleach, borax

Maintenance

Product/ Material	Use	Health and Environmental Risks	Alternatives
solvent	thin paint	nervous system damage; headaches, dizziness, coma; organ damage, respiratory damage, skin disease; death	use water- or vinyl-based paint; use paint before it thickens

TOUR YOUR SCHOOL

With the help of one or more school staff members, students will learn about chemicals in relation to your school's current supplies and maintenance practices and will then tour campus sites where such materials are used.

Objectives

After completing this activity, students will be able to

- identify some of the products being used
- describe where and why the materials are being used

Prepare

Arrange for one, two, or more speakers from school maintenance, purchasing, or administration to come to your class to discuss chemicals used at school. Provide a copy of Activity Sheet 6 for them to use as a guideline. Since responsibility for materials varies from school to school and district to district, you will need to find out who is in charge of a particular topic you think would be productive for your students to explore. For example, if pesticides are a topic of choice, you might need to have someone from building maintenance come talk about what, where, and why pesticides or insecticides are used inside the school, and ask someone from groundskeeping to come talk about pesticides, herbicides, and insecticides used outdoors. After all the guests have spoken or during their presentations, students can tour sites where these materials are used and make notes and observations about the specific locations. *Please note that such tours should not include any locations where recent applications could pose a risk to students or where hazardous materials are stored.*

Some school districts provide information about hazardous materials in handouts called Material Safety Data Sheets, or MSDS's. If you have not already done so, you might request MSDS's for chemicals used at your school at this time. They can provide additional background and reference materials (when students conduct their audit later on.)

Explain that student questions and investigations are not intended to be critical of school management, policies, or practices. Rather, they are intended to make students aware of the need to handle certain materials with care, to minimize the use of certain materials, and to take responsibility for changing their own behavior in ways that will help reduce the need for products that are hazardous.

See the Focus section for information to present to students before they meet with the school staff persons or tour sites.

Materials

Activity Sheet 6

Setting the Stage

Discuss briefly the questions on student page 30.

- **What chemicals are being used?**

  Responses might include pesticides, herbicides, insecticides, cleaning and disinfecting products, window cleaners, floor waxes, paints and solvents, and so on.

- **Where and why are the chemicals being used?**

  Answers will vary.

Focus

1. Before school staff members come to class, explain that students will conduct a detailed analysis of the use of chemical products in their own environment in later activities. They will examine what products are used for what purposes; where they are used; and what safer, healthier alternatives to their use might exist. Suggest that in this activity, the alternatives students identify will probably be specifically related to their site observations, although any additional ideas or insights they can generate based on background from the Explore activities, speakers' comments, or elsewhere will be useful. The information they gather and the observations they make will help them plan and conduct their school audit later on.

2. Distribute Activity Sheet 6 to students, and encourage them to use the concept clusters to make notes and jot down ideas and observations that occur to them during the presentations and tours. Remind students to use a separate cluster for each type of hazardous material. Explain that if there are more than four materials, they can create additional concept clusters in their Journals and record their notes there. Students should ask questions such as those on the student page to gain information that will help them complete parts of the concept clusters. The tours should provide additional information, especially information about a particular site.

It's a Wrap

Review Activity Sheet 6. Have students share their findings and add to their notes. Have students save their completed activity sheets and Journal notes to refer to as they plan and conduct the school audit.

If students do not have time to write paragraphs, have volunteers describe how the talks and tours helped them find out how and why chemical products that can be hazardous are used at their school.

Home

Follow-Up Discuss some of the common chemical products that students found in various places in their homes.

Assignment Home activity on student page 31.

Journal: Identify one cleaning product used in your home. Using appropriate care, look at the label. Is it a toxic product? How do you know? Record your findings in your Journal.

ACTIVITY SHEET

ANALYZE

Name

SCHOOL CHEMICAL PRODUCTS USE (part 1)

On the notecards below, record ideas and information about hazardous materials used at your school. Use a different card for each type of material.

Answers will vary.

Material
What is it?

Why used?

Where used?

Hazards?

Site observations:

Are there other ways to deal with the problem or address the need?

Material
What is it?

Why used?

Where used?

Hazards?

Site observations:

Are there other ways to deal with the problem or address the need?

ACTIVITY SHEET

Name

SCHOOL CHEMICAL PRODUCTS USE (part 2)

Material
What is it?

Why used?

Where used?

Hazards?

Site observations:

Are there other ways to deal with the problem or address the need?

Material
What is it?

Why used?

Where used?

Hazards?

Site observations:

Are there other ways to deal with the problem or address the need?

PLAN YOUR CHEMICAL USE AUDIT

7 ANALYZE

This lesson helps students assign tasks and plan how to get the information they will need to complete their assessment of current chemical use at the school.

Objectives

After completing the activity, students will be able to

- describe their Action Group's responsibilities
- identify questions about specific chemicals or chemical use and sources for answers
- suggest categories of products and/or sites that would benefit most from investigation

Prepare

Conducting an audit of chemicals used at school can pose some unique challenges, depending on the nature of products used and district guidelines. Safety issues may prevent students from having direct access to many of the products that they are investigating, but there are other ways for them to get the information. First, staff members responsible for buying or using specific products may be able to provide a list of products or copies of labels for students to use to get information. Second, as mentioned earlier, many school districts have printed information (MSDS) about some of the chemicals in question, and these are excellent resources for students. Third, students may choose to develop a survey to distribute to various staff members. Students can then use the surveys to complete some of the information required for the audit. If it is likely that students will need a survey in order to collect useful data, you may want to offer them the sample form on page 50 to use as a guide.

Students can fill out the blanks at the top of the form with the staff member's name, the person or place to return it to, and the date it is needed by. Then forms can be sorted according to the person they are directed to, and any duplications can be pulled before they are sent on. See the Focus section for when to have students complete this step.

Materials

Activity Sheet 7 for each student

Setting the Stage

Discuss briefly the questions on student page 33.

- **What are your Action Group's responsibilities?**

  Answers will vary.

- **What questions about specific chemicals or types of chemicals do you need to answer? Where will you get the information?**

  Answers will vary, but it is important for students to identify people or other resources they will use at this time.

- **What categories of products and/or sites would benefit most from investigation?**

  Students will probably cite pesticides, insecticides, and herbicides; cleaning materials; and maintenance products as product categories; some may choose to look at art supplies, building materials, and personal care products. Sites will vary.

Staff member: ______________________________ Date: ________________

Our class is investigating the use of chemicals at our school. In order to complete our work, we need to find out more about what is being used, and why, where it is being used, how often it is applied, and what the known hazards of its use are.

Please fill out the form below. Include information from the product label or information provided by the district. Return the completed form to

__________________________ by __________________________ .

Product: ____

Use ______________ Frequency of Use ______________ Site ______________

Warnings or Cautions: ____

____

Health/Environmental Risks: ____

____

____

Special Handling Requirements: ____

____

Thank you for your help on this environmental project.

Focus

1. Students first need to decide how to organize the audit—by site or by product category. Discuss possible options for four to six different focuses within each organizing structure. For example, a site structure might break down in the following way: (1) grounds, (2) eating areas (indoor and outdoor), (3) bathrooms, (4) gym and locker rooms, and (5) hallways and classrooms. A product category structure might have the following breakdown: (1) pesticides; (2) herbicides; (3) insecticides; (4) cleaning and disinfecting products; (5) window cleaning, floor waxing, and polishing products; and (6) paints, varnishes, and solvents.

2. Once students decide how to organize the audit, they can be divided into Action Groups, each with a specific focus. If you have additional resources from the district or if you have decided a questionnaire will be needed, discuss them with students and distribute them at this time.

3. Have students within Action Groups brainstorm tasks and questions related to their focus that need to be addressed in order to analyze current chemical use. Then have groups select a recorder and complete Activity Sheet 7. Emphasize the importance of filling in the How to Get Information column; suggest that students write *questionnaire* where appropriate.

It's a Wrap

Review Activity Sheet 7 with the students and discuss the tasks students recorded in their Journals and how they expect to accomplish them.

Home

Follow-Up Have students share information about and compare the cleaning products they wrote about in their Journals.

Assignment Home activity on student page 34.

Have you ever had a pet, such as a cat or a dog, that had fleas? Or have you ever had a problem with other pests in your home? Write a paragraph in your Journal about the products or procedures that were used to get rid of the pests.

ACTIVITY SHEET 7 ANALYZE

Name ______________________

CHEMICAL USE AUDIT PLAN (part 1)

Action Group: ______________________

Action Group Focus: ______________________

Member	Task	Questions	How to Get Information	Due Date
Action Groups will use this chart to outline a plan.	Take notes at meetings.			

ACTIVITY SHEET 7 ANALYZE

Name ______________________________

CHEMICAL USE AUDIT PLAN (part 2)

Member	Task	Questions	How to Get Information	Due Date

8 ANALYZE

CONDUCT YOUR CHEMICAL USE AUDIT

Students will work in Action Groups to assess present practices and use.

Objectives

After completing this activity, students will be able to

- survey information about chemical product use and hazards
- summarize information about how chemical products are used and the kinds of hazards they pose

Prepare

Review the task lists completed by each Action Group in Activity 7. Suggest ways that students can avoid duplicating each other's efforts. If the class sent out questionnaires, organize those that have been returned to distribute them to the appropriate Action Groups. See the Focus section for information on when to distribute them. Be on the lookout for and help students troubleshoot any areas in which they are having difficulty getting the information they need. Encourage them to scan the Issues and Information sections in the back of the book to find information they cannot locate elsewhere.

Make school administrators and others aware of what students will be doing as they visit various sites at school to conduct the audit.

Materials

Activity Sheet 8 for each student

Setting the Stage

Discuss briefly the questions on student page 36.

- **What site or category of chemicals will you investigate?**

  Students can explain how the audit is organized and the reason they chose to organize it that way.

- **What information will you need to collect?**

  Students can describe the questions they will be trying to answer about each product during the audit.

Focus

1. Action Groups get together to brainstorm questions that will help them find out what they need to know for their assessment of chemical usage. Make sure that they understand how to use the Activity Sheet to record information. Suggest that they use the last column, Additional Information, to record any information, such as use or intended use, frequency of applications, special requirements for disposal, and so on that does not fit elsewhere. You might want to explain that the cost per unit column will be used later on to compare the costs of current products and practices with the cost of alternatives that students come up with, so it is important that students use appropriate measuring tools. For example, in the case of pesticides, students will need to find out the cost per square yard treated; whereas cleaning products generally will need to be compared by liquid volume.

2. Encourage students to develop good questions to consider when gathering information during the audit. The list in the student book provides a good start.

3. Have students look at Activity Sheet 8 and their list of questions to help them figure out how to get started. Make sure that they understand how to fill in answers to their questions. Encourage them to refer to Activity Sheet 7 as a reminder for how and where to get information. Distribute the questionnaires that have been returned to the appropriate Action Groups. If there are questionnaires still outstanding, students can send out reminders at this point.

4. Students working in their Action Groups can complete Activity Sheet 8 as they conduct the site or product category audit. Whichever way the audit is organized—whether by site or product category—students will need to observe site conditions and gather information about products in order to fill in the sheet.

It's a Wrap

Review Activity Sheet 8 and monitor student progress.

Have students share their insights and observations about the use of chemical products at the school.

Home

Follow-Up Ask volunteers to discuss some of the products and procedures used to get rid of fleas or other pests in their homes.

Assignment Home activity on student page 37.

In your Journal, describe ways in which hazardous chemical products in your home could be used more safely or less frequently.

ACTIVITY SHEET

ANALYZE

Name ______________________ Action Group ______________________

CHEMICAL USE AUDIT FORM

Use the chart to record the information you gather during your assessment of chemical use.

Product	**Site**	**Cost**	**Hazards/Label Warning**	**Additional Information** (intended use, frequency, etc.)
Students working in Action Groups will use the form to record information.				

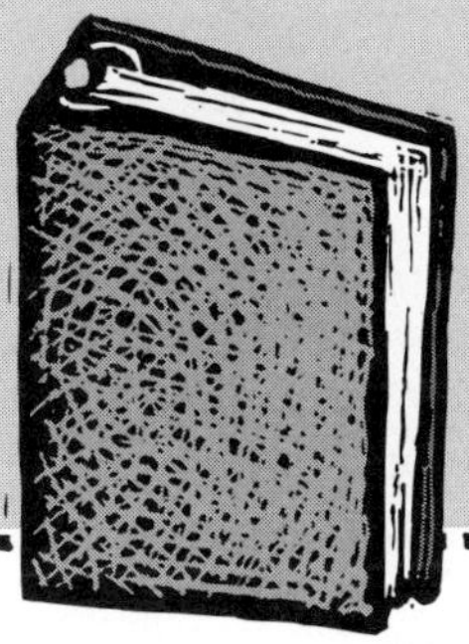

SUMMARIZE FINDINGS

Now that all the chemical use assessments have been completed, each Action Group can summarize its audit findings and compile them into a table showing pros and cons of various products and practices.

Objectives

After completing this activity, students will be able to

- summarize what they have learned about the use of chemical products in the school
- identify the pros and cons (both monetary and nonmonetary) of current use
- present the information so that it can be easily understood and interpreted

Prepare

If possible, provide reports from school committees, city council proceedings, or other similar settings to show how information gathered from different sources can be organized and presented. Students can use these resources as well as information from Material Safety Data Sheets to stimulate ideas about how to put together their final reports.

Materials

Activity Sheet 9 for each student

Setting the Stage

Discuss briefly the questions on student page 39.

- **What have you learned about the use of chemical products in your school?**

  Students may share information as well as their concerns about products and levels of use.

- **What are the pros and cons (both monetary and nonmonetary) of current use?**

  Responses include price; effectiveness; health and environmental hazards; ease or difficulty of use, storage, and disposal; and so on.

- **How can you present the information so that it can be easily understood and interpreted?**

  Action Group summaries will demonstrate how to use information to make a clear and useful presentation.

Focus

1. Have students meet in their Action Groups. Explain that they will use Activity Sheet 9 to combine information they gathered on Activity Sheet 8 with other research and observations. This summary will be the basis of a report they will write to present their findings and will be useful in the next section, Consider Options, when they research and analyze alternatives to current chemical use.

2. Working in their Action Groups, students will brainstorm the costs and benefits of specific chemical products. Encourage them to look at the questions on the student page to spark ideas about how to evaluate different aspects of a product and the pros and cons of its use.

3. After students have filled in the activity sheet with as many pros and cons as they can think of for each product, discuss their ideas. Allow them to change or add to their sheets according to what comes up in discussion.

4. Students get together in Action Groups again to write a one-page report describing the results of their Chemical Products Audit. Have students present their reports to the class and discuss differences or similarities in them.

It's a Wrap

Discuss students' answers to It's a Wrap questions.

- **In your opinion, which chemical product used at your school looks worst when you weigh pros and cons?**
- **Which chemical product looks best?**
- **What surprised you the most about your summary?**

Home

Follow-Up Ask students to share what changes they would make in the purchase or use of chemical products in their homes.

Assignment Home activity on student page 40.

Based on what you found in your Chemical Use Summary, is there a chemical product in your home that you would like to get rid of? In your Journal, describe what the product is, what it has been used for in your home, why you would like to get rid of it, and how you propose to dispose of it.

ACTIVITY SHEET 9 ANALYZE

Name ____________________

CHEMICAL USE SUMMARY

Complete the table below to summarize your findings.

Product/Use	Pros	Cons
Students in Action Groups will use this chart to compile results of the Chemical Use Audit and prepare a summary.		

ACT LOCALLY

Students can use the following outreach activities to apply their learning, share information, and strengthen their ties to the community.

Now that you have a better understanding of how chemical products are used at your school, find ways to share what you have learned. Think of a project that will help increase awareness of some of the hazards of chemical products, or use one of the suggestions given below.

1. Enhance awareness of the importance of choosing chemical products wisely by organizing a contest to select the best chemical products awareness slogan. Award prizes, such as t-shirts, books, posters, or bumper stickers printed with the slogan, to the students with the top entries.

2. Ask the school administration to establish a "Clean and Green" Awareness Day. Invite resource groups in your area to bring informational materials and set up tables in the cafeteria or main entrance to your school. Most local groups will be happy to help you with this project—many have a staff person whose job is community outreach. If a group is unable to attend, ask if someone can send information and resource materials to be displayed.

3. To raise awareness of the hazards of throwing toxic chemicals into the trash or pouring them down the sink, make posters to display around school. Consider drawing a flowchart showing how chemicals get into our water and soil, making a list of ten commonly used products that contain chemicals that need to be disposed of in special ways—or come up with ideas of your own.

Consider Options

BRAINSTORM ALTERNATIVES

10 CONSIDER OPTIONS

Students will use what they have learned to target five to six chemical products that they would like to see eliminated or used less in the school. Then they will brainstorm ideas for alternatives to these harmful chemicals that they have already have learned about and research to find additional alternatives. The alternatives may include products, changes in behavior, attitudes, or practices, and physical alterations to the school environment.

Objectives

After completing this activity, students will be able to

- identify chemical products that are most harmful
- list a variety of alternatives to reduce or eliminate the use of harmful chemical products at the school

Prepare

Students will need information about alternative products and technologies, practices, attitudes, and so on. Much of this has been introduced in prior activities or is contained in the Issues and Information sections in the back of the student book. However, depending on the chemical products students choose to focus on in this lesson, there may be gaps in the information about alternatives. You may wish to get additional printed materials from the library or from various environmental organizations to make available to students as they work. See the Focus section below for when to introduce these materials to the class.

Before students meet in their Action Groups for their brainstorming, remind them that alternatives to harmful chemical products fall into three categories: using different, less toxic products; changing behavior, attitudes, or expectations so that fewer chemicals are used; and physically modifying the environment so that fewer chemicals are needed. Suggest that, in some cases, a combination of these alternatives provides the best solution.

Materials

Activity Sheet 10 for each student (See Blackline Masters section.)

Setting the Stage

Discuss briefly the questions on student page 45.

- **What chemical products used at school do you think are most harmful?**

  Answers will depend on the specific information students have received or discovered, but will likely be clustered in one or more of the three categories that have been investigated: pesticides, insecticides, and herbicides; cleaning products; and maintenance products.

- **What are some of the ways to effectively reduce or eliminate the use of harmful chemical products at your school?**

  Possible responses include changing products, changing behaviors or attitudes, and altering the physical environment.

Focus

1. Launch the discussion by inviting volunteers to list some of the chemical products they investigated that concerned them most and why. Then have students work in their Action Groups to identify five or six chemical products that they all agree they would like to focus on. Remind them that they can use the information on Activity Sheet 9 to help them decide which products to choose.

2. For each product they have selected, have students brainstorm alternatives, using information they already have and questions on the student page to stimulate ideas about additional areas of research.

3. Make sure that students are able to find a source for information they need to pin down suitable alternatives. You may want to remind them of the resources in the Issues and Information sections in the back of their books, of the notes they took on Activity Sheets 5 and 6, and of other reference materials you can make available to them at this time (including materials suggested under Teaching Environmental ACTION, page 8 of this book).

4. Students can use Activity Sheet 10 to record alternatives to the products they've targeted. Suggest that they consider multiple alternatives or a combined approach to the products they have targeted. Remind them to use their Journals as needed to create additional concept clusters.

It's a Wrap

Invite students to share their ideas about harmful chemical products and their alternatives. Help students identify connections between products with similar alternatives.

Home

Follow-Up Have students discuss and compare their responses to the questions about which chemicals they would like to get rid of, why, and how. If there is a hazardous waste disposal site or day in your community, make sure that the information comes up in discussion. If possible, have volunteers contact the local waste disposal company or agency to get information about their rules for chemical products and their disposal. Discuss the importance of not just throwing things down the sink or into the landfill by focusing on the specific impact in your community—polluting salmon streams, contaminating precious aquifers, and so on.

Assignment Home activity on student page 46.

Journal

In your Journal, record ways in which you could change behavior, habits, or the physical environment of your home to reduce your need for or use of chemical products. What would the changes be? What chemical products would the changes impact?

Name

ALTERNATIVES TO HARMFUL CHEMICALS (part 1)

Use the notecards on these sheets to organize your ideas about alternatives to harmful chemicals. Fill in a different card for each product.

Product

Answers will vary.

Alternative Product

Modifications to School Environment

Behavior/Attitude Change

Product

Alternative Product

Modifications to School Environment

Behavior/Attitude Change

Product

Alternative Product

Modifications to School Environment

Behavior/Attitude Change

ACTIVITY SHEET

Name ____________________

ALTERNATIVES TO HARMFUL CHEMICALS (part 2)

Product

Alternative Product

Modifications to School Environment

Behavior/Attitude Change

Product

Alternative Product

Modifications to School Environment

Behavior/Attitude Change

Product

Alternative Product

Modifications to School Environment

Behavior/Attitude Change

WEIGH COSTS AND BENEFITS OF ALTERNATIVES

11 CONSIDER OPTIONS

Students will evaluate the alternatives to using harmful chemical products that they generated in Activity 10, weighing costs and benefits. Both monetary and nonmonetary costs will be addressed, as well as short-term and long-term benefits.

Objectives

After completing this activity, students will be able to

- identify the benefits of an alternative
- identify the monetary and nonmonetary costs of an alternative

Prepare

Before students begin trying to assess the costs and benefits of a particular alternative, make sure that they have a way to find the actual price of products. Prices for some cleaning products can be found in Issues and Information section B; other prices can be found by looking at specific items in a grocery or hardware store or by calling a store for price information.

Discuss the questions on the student page. Remind students that the cost of something may not just be its price, there may be costs related to human health and the environment.

Materials

Activity Sheet 11 for each student

Setting the Stage

Discuss briefly the questions on student page 48.

- **What factors need to be considered when assessing the benefits of an alternative?**

  Responses should reflect an understanding of both direct and indirect benefits.

- **What factors need to be considered when assessing the costs (both monetary and nonmonetary) of an alternative?**

  Responses may include issues of relative effectiveness, the difficulty of changing behaviors, and other more subtle kinds of costs in addition to direct costs.

Focus

1. Before students begin, walk them through a sample alternative, using the questions in the student book to guide the discussion. For example, if the suggestion was to switch from using drain cleaner to using a plumber's snake, the discussion might go as follows:

- **What is the monetary cost (price) of the change?**

  purchasing the snake

- **What are the nonmonetary costs?**

  might have to train people to use it

- **Are there any hidden costs? What are they?**

  might take up more storage space; may have to be disinfected, perhaps with chemicals

- **Are changes in behavior or attitude required?**

  staff might consider it "yucky"

- **If so, what are the "costs" involved?**

  re-educate to inform staff of how much healthier it is for them and for the water

- **Is the alternative as effective at doing the job? Is this a cost or a benefit?**

  probably; benefit

- **What are the short-term benefits to health and the environment?**

  A snake avoids the hazards of drain cleaner, which is caustic, can harm skin and eyes, is harmful or fatal if swallowed, and has hazardous fumes.

- **What are the long-term benefits to health and the environment?**

  Because a snake is mechanical rather than chemical, it does not cause water pollution as does drain cleaner.

- **Are there other benefits?**

  Drain cleaner is a corrosive; it can eat through pipes and make expensive plumbing repairs necessary.

2. Students can work on their own or with a partner to evaluate their Action Group's alternatives to harmful chemical products and record their findings on Activity Sheet 11.

It's a Wrap

Have students share with their Action Group what they discovered about the costs and benefits of the alternatives to harmful chemical products. Encourage students to add new ideas to their sheets as they are prompted by the discussion. Have them use their Journals to note one unexpected cost and one unexpected benefit that they discovered.

Home

Follow-Up Have students discuss ways they could change behavior, habits, or the physical environment in their homes to reduce the need for or use of chemical products. What would the changes be? What chemical products would the changes impact? How would they get the cooperation of family members in the effort?

Assignment Home activity on student page 49.

Suggest one change to your family in the way chemical products are used at your house. What are the costs and benefits of the change? In your Journal, make a table similar to the one on Activity Sheet 11 to record the information. Then evaluate your idea. Does the change make sense? Why?

ACTIVITY SHEET

CONSIDER OPTIONS 11

Name

COSTS/BENEFITS ANALYSIS OF ALTERNATIVES

Complete the table below to record your ideas.

Product/Use	Alternative	Costs	Benefits
Answers will vary but should include a variety of costs and benefits, both direct and indirect.			

MAKE RECOMMENDATIONS

12 CONSIDER OPTIONS

Based on their analyses and evaluations, students choose which alternatives they will recommend to the class and plan how to make a convincing presentation. Their presentation plans will include specific costs and benefits information and provide graphics or other visuals to support their recommendations.

Objectives

After completing this activity, students will be able to

- identify criteria for selecting which alternatives to recommend
- list information that they will want to include in their presentations

Prepare

You will need to provide tabletop and bulletin board space in the classroom for Action Groups to set up their displays, as well as graph paper and art materials for them to use to make them.

Materials

- Activity Sheet 12
- graph paper
- art supplies
- oak tag or poster board

Setting the Stage

Discuss briefly the questions on student page 51.

- **What are the most important considerations to keep in mind as you decide which alternatives to chemical products used at the school you will recommend?**

  Responses may include benefits to health and the environment, ease of implementation, and cost.

- **What are some kinds of information that you could use to support your recommendations?**

  Specific information about health and environmental hazards of current products and practices, cost comparisons, charts, graphs, and tables that represent the data, and so on.

Focus

1. Before students decide on which recommendations to make to the class, have each Action Group meet to review the cost and benefit evaluations that pairs of students completed in Activity 11. Remind students that no alternatives are ideal in every way; there are trade-offs for every idea. Explain that health and environmental benefits, while they may seem harder to quantify, can and should be emphasized for the presentations. Students may find ideas in the Issues and Information sections in the back of their book to help support them in this regard.

2. When students have identified the alternatives they want to recommend to the class, have them use Activity Sheet 12 to plan their presentation. Encourage them to use the back of the sheet to produce rough sketches or graphs or other visuals they may want to create for their presentation.

3. Make graph paper and art materials available so that students can use them to create visuals and make their displays. Encourage students to incorporate information from Activity Sheet 12 into their display. Allow time for students to collect or generate visuals and graphics and put together their displays.

It's a Wrap

Each group should set up a display in the classroom in preparation for their presentation, which will be made in the next activity. Then bring the groups together and discuss alternatives that were rejected and the factors that led to their elimination. Encourage groups to identify what it would take to implement rejected ideas or what might be changed to make an idea more feasible.

Home

Follow-Up Discuss the costs and benefits analyses students made at home. Did the changes students investigated make sense? Why or why not?

Assignment Home activity on student page 52.

Think about the alternative to chemical use at home that you evaluated. In your Journal, write a plan for presenting the idea to your family. Your plan should take into consideration issues such as who would be responsible for purchasing the alternative or for learning about how to use it, who would benefit from it and how, and other aspects such as cost, disposal, and so on.

ACTIVITY SHEET 12 CONSIDER OPTIONS

Name ______ Action Group ______

PRESENTATION PLAN (part 1)

Use this sheet to organize the presentation your Action Group will give to the class when you recommend your alternatives.

Alternative ______ Current ______

Costs **Responses will include price and other costs, such as retraining.**

Benefits **Responses will emphasize health and environmental benefits.**

Implementation plan **Responses will include attitude changes or modifications to the school required.**

Alternative ______ Current ______

Costs ______

Benefits ______

Implementation plan ______

Alternative ______ Current ______

Costs ______

Benefits ______

Implementation plan ______

ACTIVITY SHEET 12 CONSIDER OPTIONS

Name ________ Action Group ________

PRESENTATION PLAN (part 2)

List of Visuals You Will Use to Support Your Recommendations

Description	Source	Person Responsible
Students may list graphics		
from sources they have found		
or they may choose to create		
their own.		

Use the space to sketch out a table, chart, or graph that you want to use to show some of your information.

Students might sketch some sort of matrix comparing health and environmental costs of current and alternative products or processes, or any other graphic they choose. Other visuals might include pictures showing chemical burns, drawings of how a particular toxin concentrates in the food chain, and so on.

ACT LOCALLY

Students can use the following outreach activities to apply their learning, share information, and strengthen their ties to the community.

Now that you have explored a wide range of alternatives to harmful chemical products, decide on an action project that your class can undertake now. Use one of the ideas below, or think up one of your own.

1. Create a one- or two-page list of supplies and maintenance ideas and resources to have printed in the school newspaper or as a flyer to distribute to students.

2. Set up a display of nontoxic cleaning, maintenance, or pest control products to show other students and teachers alternative products and practices. To avoid having to purchase products, try using empty containers or writing to manufacturers to get (free) promotional products. If these are unavailable, make up your own containers with labels based on real products.

3. Participate in an event—a hike, fund-raiser, clean-up activity—organized by an environmental group in your community.

4. Make a presentation about this project to your school's Parent-Teacher organization. This group may be able to support some of your alternative suggestions.

Take Action

CHOOSE ALTERNATIVES

Action Groups will present their proposals for alternatives and the class will reach a consensus about which proposals to present to school officials.

Objectives

After completing this activity, students will be able to

- evaluate the effectiveness of presentations
- evaluate and compare alternatives to chemical use
- reach a consensus with classmates

Prepare

Students will fill out Activity Sheet 13, ranking on a scale of 1 to 3 features of each alternative presented. You may wish to make a transparency of Activity Sheet 13 to display on the overhead projector; rankings can be compared, tallied, and averaged in order to reach a consensus.

Materials

- Activity Sheet 13 (Each student will need one copy of the sheet for each alternative measure being presented. See Blackline Masters section.)
- Transparency of Activity Sheet 13 and overhead projector for tallying student responses (optional)

Setting the Stage

Discuss briefly the questions on student page 57.

- **How can you evaluate the merits of a presentation?**

  Responses might include the quality of arguments in favor of an alternative, the effectiveness of graphics and other visuals to support the alternative, and so on.

- **What factors will help make implementing an alternative possible?**

  Factors may include low cost, high benefits, ease of implementation, high level of acceptance of the alternative, and so on.

- **What factors will help make an alternative successful?**

  Responses may include the effectiveness of the alternative, its acceptance by students and staff, its appropriateness to the environment, and so on.

Focus

1. As each Action Group presents its recommendations, students can use Activity Sheet 13 to rate the presentation. Suggest to students that they evaluate the effectiveness of the presentation on how well the recommendations were supported by the facts presented, the use of graphics and other visuals to support the recommendation, and so on.

2. Tally the student responses so that there is a class ranking sheet for each measure. Encourage discussion as the rankings are interpreted and compiled, taking into consideration questions such as the following:

- **Is there one kind of chemical that the school uses more than others?**
- **Is there a chemical that has a more serious impact on health and the environment and therefore would have the most impact on health and the environment if it were replaced?**

- **Does one recommendation involve an investment that will pay for itself quickly in savings?**

3. After students have prioritized the recommendations, help them to reach a consensus on which measures to propose to the school for implementation.

It's a Wrap

Discuss students' impressions of and opinions about the alternative products and measures the class decided on. Invite students to share their slogans and cartoons.

Home

Follow-Up Students can share the alternatives they plan to recommend to their families, their strategies for presenting their ideas, and the factors that influenced their decisions.

Assignment Home activity on student page 58.

Present your alternatives to chemical use to your family. Have family members rate your idea according to how much it would reduce the health and environmental risks to your family, how much time it would take, how each family member could participate, how much it might cost, and so on. Have them use a ranking system similar to the one on Activity Sheet 13 to reach a consensus about which measures to adopt.

ACTIVITY SHEET 13 TAKE ACTION

Name ______________________________

RATING SHEET

Fill in the following rating sheet for each presentation.

Group ______________________________

Plan ______________________________

Costs

Expensive • • • Inexpensive

Health and Environmental Benefits

Low • • • High

Long-Term Effectiveness

Low • • • High

Difficulty of Implementing

Low • • • High

Cooperation Incentives

Low • • • High

Effectiveness of Presentation

Low • • • High

Additional Factors to Consider

____

____

____

Priority

Low • • • High

PREPARE PROPOSAL

14 TAKE ACTION

Students will write a proposal to the school administration or appropriate body outlining costs, benefits, and ways to implement alternatives to chemical products and processes in current use.

Objectives

After completing this activity, students will be able to

- assign tasks to complete a project
- use an outline to prepare a proposal
- support recommendations with sound arguments and supportive materials

Prepare

Students will be preparing a proposal to submit to the school. You will need to identify the individual or group to whom to direct the completed proposal (if you have not already done so). Before the students begin writing, you may want to discuss the audience that you have identified and how it might influence students' work. (See Focus, below.)

Materials

- Activity Sheet 14
- art materials

Setting the Stage

Discuss briefly the questions on student page 60.

- **What chemical use problems were discovered during your audit?**

  Answers will vary.

- **How do the alternatives you chose offer short- and long-term solutions?**

  Responses should include health and environmental benefits, costs, and other factors.

- **What main factors led you to choose these particular alternatives?**

  Responses should indicate that students believe the alternatives will be both beneficial and effective, are likely to have long-term success, and have high probability of being implemented.

Focus

1. Discuss the committee or school officials receiving the proposal to help students focus on addressing a specific audience. For each measure being proposed, brainstorm ideas about how to present the measure. Encourage students to think about questions such as the following:

- **What do you want to emphasize?**

  Students should recognize that the health and environmental benefits of the alternatives will probably be their best selling points. How can they make those benefits clearest to the target audience? What kinds of supporting data do they need to include in the proposal that will educate the audience about the hazards of current use?

- **How can charts, graphs, tables, and other visuals help you illustrate your ideas and provide data to support your recommendations?**

  Students can suggest ways to display information and convey their message.

- **What tone do you think will be most appropriate and persuasive?**

  Students can determine what approach will be most diplomatic and effective with the audience they are trying to persuade. If the audience is already aware of chemical issues and concerns, the approach students take will be different from what it would be for an audience that does not have as much background. In addition, you might want to write the following statements on the chalkboard and discuss how each comes across:

These chemicals are making us all sick; if you switched to alternatives, the absentee rate would go way down.

The research shows that short-term reactions to pesticide exposure can include nausea, headaches, shortness of breath, and an inability to concentrate.

2. Have students work together on an outline for their presentation, using Activity Sheet 14 to keep track of assignments and progress.

3. Invite students to suggest additions or revisions to the outline on the activity sheet before they begin. Allow time for them to revisit the outline periodically to monitor progress on the various parts of the proposal.

It's a Wrap

Students can put together the elements of their proposal and revise parts of it as necessary before delivering or presenting it to the appropriate group.

Reminder At this time you may wish to administer the Content Quiz and Student Survey. Your students may already have taken these as pretests. They may now be given as post-tests. In addition, the Student Self-Evaluation Form can now be used to help students assess their own progress. These forms are all included in the Blackline Masters/Assessment Tools section of your Resource Guide. Answers to the Content Quiz are on page 16.

Home

Follow-Up Have students share how family members reacted to and rated their proposals for alternatives to chemical use at home and which proposals families agreed to adopt.

Assignment Home activity on student page 61.

Write a letter about chemical use to a city official, congressional representative, or even the President. Tell the person what you like about their health and environmental policies or give specific arguments to encourage better decisions regarding chemicals. Make a copy of your letter for your Journal.

ACTIVITY SHEET 14 TAKE ACTION

Name

PROPOSAL CHECKLIST (part 1)

Use this checklist to plan and monitor tasks that may need to be done in order to complete your proposal. You can add to the list as needed.

Tasks	Person Responsible	Due Date
1. TITLE ☐ Cover illustration ☐ Proposal statement		
2. WRITE THE INTRODUCTORY PARAGRAPH. ☐ Explain the project. ☐ Briefly describe audit findings.		
3. WRITE RECOMMENDATIONS. ☐ Outline each plan. ☐ Highlight the benefits. ☐ Specify the costs. ☐ Suggest a step-by-step plan for implementation. ☐ List ideas for increasing awareness, changing attitudes, and encouraging support. ☐ Pinpoint projected savings.		

ACTIVITY SHEET

Name

PROPOSAL CHECKLIST (part 2)

Tasks	Person Responsible	Due Date
4. PRESENT RESEARCH FINDINGS. ☐ Prepare graphs, if needed. ☐ Design tables or charts. ☐ Prepare illustrations, photographs, or other art work.		
5. WRITE CLOSING STATEMENT. ☐ Summarize briefly what steps you have recommended and why.		
6. MAKE A BIBLIOGRAPHY. ☐ List the resources that you used to generate your recommendations.		

FOLLOW THROUGH

TAKE ACTION 15

Students follow up on the proposal as it is implemented on campus and suggest ways to increase implementation of remaining measures.

Objectives

After completing this activity, students will be able to

- summarize the status of implementation of the measures they proposed to the school
- identify areas in which additional effort or education needs to occur in order for implementation to proceed

Prepare

Review the school's response to the alternatives proposed by students. Establish a way to track success and continue advocacy of the measures proposed. Activity Sheet 15 can be used as a guideline.

Materials

Activity Sheet 15

Setting the Stage

Discuss briefly the questions on student page 63.

- **How can you find out the effects of your proposal?**

  Responses will likely focus on how widely proposals have been or are scheduled to be implemented.

- **How can you increase implementation of your recommendations and raise awareness of the need for change?**

  Responses include educating students and staff to the hazards of certain chemicals and working to change attitudes and expectations, for example, teaching tolerance for insects that don't pose a health risk; acceptance of floors that aren't highly polished, and so on.

Focus

1. Discuss the proposal students submitted and the response it received. Help students analyze what they can do to improve implementation of their ideas. Have them use Activity Sheet 15 to evaluate the status of their recommendations, keep track of implementation, and list ideas.

2. After one or two months, have students fill in the activity sheet and check to see if any additional changes have been implemented. If changes have still not been made, encourage students to try to determine whether the changes are scheduled for a later date or if the idea has been tabled for a particular reason. Have them continue to follow up, looking at their successes and what still could be done.

It's a Wrap

Discuss students' responses to It's a Wrap questions and the ongoing success of their proposals.

Home

Follow-Up Have students share the letters they wrote. Invite students to share the responses they receive, as well, when they arrive.

Assignment Home activity on student page 64.

Use your Journal to write about the impact that your changes in chemical use are having on your household. Explain how the alternatives are working, and how family members are reacting to the changes.

ACTIVITY SHEET

TAKE ACTION 15

Name ______________________

IMPLEMENTATION TRACKING SHEET

Use this tracking sheet to evaluate the status of each of your recommendations.

Proposal	Results to Date	Ideas for Follow-Up
Students will use this sheet to follow up on their proposal.		

Additional Ideas for Increasing Implementation

Answers will vary.

Appendices

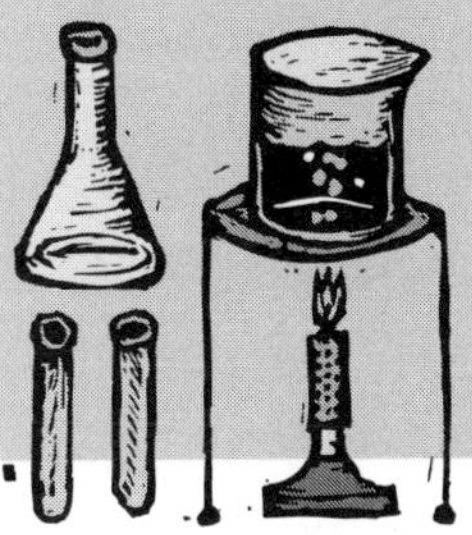

Section A

CLASSIFICATION OF HAZARDOUS CHEMICALS

We live in a world full of synthetic chemicals—detergents, paints, household cleaners, gasoline, medicines, food additives, cosmetics, art supplies, plastics. The list goes on and on. More than 70,000 synthetic chemicals are in use today, and about 1000 more are added every year. Not all of these chemicals are hazardous, but many are. *Chemicals: Choosing Wisely* focuses on hazardous pesticides, cleaning products, and solvents, but thousands of other common products also present hazards to the environment and to human health.

Unfortunately, information about how to classify chemicals as being hazardous or not is confusing and incomplete. The U.S. government has only begun to require information about the hazards of chemicals in recent years and now has a number of laws that regulate these chemicals. But the majority of chemicals have not yet been completely tested.

Classified by Characteristics

Chemicals can be hazardous in a number of different ways. Scientists classify chemicals as hazardous if they have one or more of these characteristics:

- Corrosive—they can destroy, eat away, or burn materials with which they come in contact.
- Explosive—they can explode if heated, put under pressure, shocked, or mixed with another substance.
- Flammable—they can be set on fire easily and can burn rapidly.
- Toxic or Poisonous—they can cause injury or death if swallowed or absorbed through the skin or through ingestion or inhalation.

Classified by Effects

Another way to classify hazardous chemicals is by their effects (see Issues and Information sections D, E, and F for more information on health and environmental effects of hazardous chemicals). A hazardous chemical can have one or more of these effects:

- Acute toxicity—cause sudden illness or death
- Attack on nerve cells—cause damage to nervous systems; the effect can be long-term or fatal
- Bioaccumulation—accumulate in food chains in the environment or in parts of the body, multiplying the damaging effects
- Birth defects—cause damage to babies before they are born
- Cancer—cause cancer in humans or other animals; called carcinogens
- Chronic toxicity—cause illness or death after longer exposure, often in small doses
- Environmental damage—cause destruction of habitat and wildlife
- Gene mutations—alter genes in humans or other species
- Reproductive problems—cause reproductive difficulties including infertility and premature births
- Persistence—do not break down in the environment or the body and therefore their effects persist for many years and may bioaccumulate

Labels

Federal law requires products that are known to be hazardous to be labeled. Many labels, however, warn of the effects if swallowed and may not mention the problems caused by inhalation or skin contact. The signal words on labels indicate the level of the product's hazard:

- "Danger" means highly toxic.
- "Warning" means moderately hazardous.
- "Caution" means hazardous, but less so than the other categories.

Some products are also marked "Poison" and have a skull and crossbones to indicate strong toxicity.

Difficulties in Identifying and Classifying Hazardous Chemicals

Both scientists and consumers face many difficulties in identifying hazardous chemicals. Here are some of the problems faced when trying to figure out which chemicals are hazardous and what hazards they present:

1. There are different definitions for "hazardous" and "toxic." Some scientists, for example, define toxic chemicals as only those that are fatal to humans in small doses. Some define toxic chemicals as those that cause any harm or injury to the environment or human health.
2. Many tests on chemicals measure only their acute effects—the poisoning that occurs soon after a single dose or exposure. However, many chemicals have serious chronic effects—the poisoning from even small doses or exposures repeated over days, weeks, months, or years. These chronic effects are complex to research and measure.
3. The bioaccumulation of chemicals can greatly magnify their hazardous effects, which may seem much less harmful before bioaccumulation.
4. Many chemicals become hazards only when they are mixed with other chemicals in the body or in the environment.
5. Tests on chemicals are generally done on laboratory animals. The results may or may not be the same for humans. Furthermore, children react differently to chemicals than do adults, and tests often do not measure this difference.
6. Product testing is often done by the product's manufacturer rather than by an independent researcher.
7. While laws require that chemicals be identified on a product's label, only the "active" ingredients must be listed (for example, in bug sprays, the active ingredient is the one that kills the bugs). Most chemicals are composed of many chemicals in addition to the active ingredients—and many of these unlisted ingredients, called "inert" ingredients, are hazardous and represent threats to human health and the environment.
8. Very few government resources are dedicated to investigating hazardous chemicals; even fewer are assigned to investigating the unlisted, inert ingredients.

Citizens' groups and the government are working to improve the identification and regulation of chemicals that are hazardous to human health or the environment. In the meantime, we can all help by taking action now. We can examine our homes, schools, and workplaces for hazardous chemicals and, where possible, replace them with safe alternatives. We can use proper precautions if we do use hazardous chemicals. And we can dispose of hazardous substances responsibly by never putting them in the trash, down the drain, or down storm drains.

Section B
CLEANING PRODUCTS AND SAFER ALTERNATIVES

In most homes and schools, a great many cleaning products are used. Many of these are very effective at their cleaning jobs, but the price of using them is often environmental pollution and human health problems.

Commercial Cleaning Products and their Effects

Some cleaning products can cause air pollution; water pollution; accidental poisonings (especially of children); damage to people's skin, lungs, and eyes; and long-term illnesses such as cancer. They enter the environment as gases emitted when we manufacture or use them, as water pollutants when we pour them down drains, and as land and water pollutants when we throw them or their containers away. Some researchers estimate that about 25 gallons of hazardous chemicals are used in the average American home each year.

It is difficult to know which cleaning products cause which effects. First, different products, even those for the same purpose, have very different ingredients. Some are, therefore, much safer than others for human and environmental health. Second, although more information about the effects of the ingredients is now required on packaging, some of the information is helpful, and some of it is deceiving. "Biodegradable," for instance, is a much-overused term that does not necessarily mean that a product is environmentally harmless. And third, many ingredients have not been tested thoroughly enough to be sure of their effects. With all of that uncertainty in mind, here is a list to give you an idea of some of the problems that are caused by some household cleaners:

Aerosols Some cleaners come in aerosol spray cans. Aerosols contain chemicals that contribute to air pollution; many scientists believe they contribute to the depletion of the ozone layer in the earth's atmosphere. Some aerosols have been banned by the government.

Air fresheners Most commercial air fresheners and deodorants release toxic gases into the air. They do not, in fact, freshen the air; instead they deaden the nerves in our noses so that our sense of smell is blocked. They may also contain pesticides that collect in the fatty tissues of our bodies or chemicals that affect the central nervous system.

All-purpose cleaners Ammonia in many cleaners is harmful to the lungs (ammonia is in most glass cleaners as well as all-purpose cleaners). Additional chemicals in some all-purpose cleaners are suspected of forming substances in water that can cause diseases such as cancer, birth defects, and liver damage.

Dishwashing detergents Some dishwashing detergents include an ingredient that affects the central nervous system, another ingredient that may cause liver damage, and other toxic ingredients.

Floor and furniture polishes Most of these contain an ingredient that causes cancer in laboratory animals.

Household bleach Many bleaches contain chlorine that forms substances in water that are suspected of damaging the environment and causing diseases, including cancer. In a septic tank, chlorine kills bacteria that are useful in decomposing sewage. Chlorine is also an irritant to the nose, skin, and eyes.

Laundry detergents Phosphates in detergents cause excess growth of algae in water, which causes fish, plants, and whole aquatic systems to die (some states have banned detergents with phosphates because of their devastation of lakes and waterways). Other chemicals in some laundry detergents can cause additional environmentally harmful water pollution and may cause cancer.

Safer Alternatives

The good news is that there are safe alternatives to hazardous household cleaners. Many people think the safe alternatives are more expensive, time-consuming, or inconvenient. The truth is they are readily available, easy to use, and cheaper than commercial cleaners. And they do not poison our air and water or threaten our health.

Here is a chart of some suggested safer alternatives and their prices compared with the prices of commercial products. Look in the Resources section for where to find more information on alternatives. And remember, these safer alternatives are still chemicals. They must be used responsibly and with caution. NEVER mix chemicals unless you know it is safe to do so.

Common Cleaners and Suggested Safer Alternatives

Commonly Used Cleaner	Price of Cleaner*	Suggested Safer Alternative	Price of Alternative*
Air freshener	$1.00 per 7 oz. bottle	Herbs, flowers; indoor plants purify air; box of baking soda in refrigerator or garbage absorbs odors.	$.69 per 16 oz. box of baking soda (lasts 3 months as deodorizer)
All-purpose cleaners	$.13 per gallon (diluted)	2 tsp. borax and 1 tsp. soap in 1 qt. of water (use plain soap, not deodorant soap); or baking soda and water.	$.09 per gallon (borax, soap, and water)
Disinfectant	$.20 per gallon (diluted)	½ c. borax in 1 gallon water	$.17 per gallon
Drain cleaner	$3.10 per treatment	¼ c. baking soda followed by ½ c. distilled white vinegar, flushed with boiling water.	$.22 per treatment
Floor and furniture polish	$2.46 per cup	1 c. lemon juice to 2 c. olive, vegetable, or mineral oil	$.90 per cup
Glass/window cleaner	$.52 per cup	Soap and water rinsed with ½ c. vinegar to 2 c. water. Dry with crumpled newspaper.	$.06 per cup

Common Cleaners and Suggested Safer Alternatives (cont.)

Commonly Used Cleaner	Price of Cleaner*	Suggested Safer Alternative	Price of Alternative*
Scouring powder	$.06 per oz.	Baking soda and hot water	$.04 per oz.
Toilet cleaner	$.80 per cup	Baking soda and toilet brush	$.17 per ¼ box
Tub and tile cleaner	$1.12 per cup	Baking soda and hot water	$.17 per ¼ box

*Average price of *least-expensive* products available in an urban grocery store in Oregon, August 1996. Many more expensive products are also available.

Section C

PESTICIDES AND INTEGRATED PEST MANAGEMENT

Pesticides

The different types of pesticides are classified by the pests they control: insecticides for insects, herbicides for plants, fungicides for fungi, and so forth. Pesticides have been important in controlling animals and fungi that carry serious diseases—including malaria, plague, and yellow fever—and in stopping pests that destroy crops. Manufacturers, especially since World War II, have created tens of thousands of synthetic chemical pesticides. Their benefits have been many, but we now know that they can also cause serious damage to the environment and to human health.

Pesticides are designed to harm and kill, and their effects are often not limited to the pest or the area on which they are used. They are carried far in the wind or through soil and water. They can seep as much as fifteen feet below the surface of the ground in two days after being applied, reaching the underground sources of drinking water. Pesticides have been found in drinking water sources in the majority of states in the United States.

When the pesticides spread, their poisonous effects reach to people and many other animals and plants. Many pesticides are known to cause health problems, and their effects are usually worse on children. Health problems identified with pesticides include stomach upsets, convulsions, liver damage, behavioral disturbances, asthma-like problems, cancer, gene damage, birth defects, and reproductive difficulties.

Scientists are concerned about pesticides in drinking water, in the air, and on the fruits and vegetables we eat. Many pesticides do not break down, but persist in the environment for years, even centuries. Their harmful effects are often magnified when their concentrations increase as they move up the food chain. Animals at the top of food chains (humans, other large mammals, large birds) absorb the greatest amounts of the chemicals and are at greatest risk for harm. (See Issues and Information section E.)

Pesticides, while very effective in some cases, have also backfired at times. Some have killed the natural predators of the pest for which they are being used, disturbing the whole balance of environmental relationships. In other cases, insects have been able to develop resistance to the insecticides used on them and to pass on this resistance to their offspring.

Integrated Pest Management

As a result of these problems with pesticides, Integrated Pest Management (IPM) is growing in popularity as a way of controlling pests without widespread attack with poisons. IPM is an approach that involves examining the whole environmental system and choosing the best combination of methods to deal with unwanted pests. Pesticides made from synthetic chemicals are considered only as a last resort, to be used only in carefully controlled areas and for limited times.

IPM means using the least-toxic methods possible. It means combining several different ways of attacking a pest problem including learning to be more tolerant of pests, physical barriers to keep pests out, and using nontoxic products. These are some of the least-toxic pest control methods:

- Keep things clean. Nothing discourages pests like finding no food. Clean under stoves and refrigerators as well as on counters and in cabinets.

- Keep things dry. Many insects are attracted to moisture. Fix any leaking pipes, keep surfaces dry.
- Put up barriers from entry. Caulking can keep insects from entering. Caulk around door and windows, and around cracks in walls or plumbing. Screens are also good at keeping insects out, especially insects that fly.
- Lay a "safe" trap. Pheromone traps, for example, are made of scent hormones from insects. They attract the insects but are harmless to people and pets.
- Use safe products. Many products are now available such as insecticidal soap and herbal repellents that are not poisonous to humans or the environment but are effective against pests.
- Weed them away. If weeds are the problem, try underlaying the area with plastic, strips of asphalt, or other barriers to weeds. Or, use nontoxic, biodegradable herbicides. Or, try pulling the weeds up by hand! Insects and other pests can sometimes be "weeded away," too, by simply removing them by hand.
- Let go of perfection. Perfect green lawns are usually dangerously full of toxic chemicals. Perfect fruits and vegetables are often grown with extensive use of pesticides.

IPM is an orderly process:

1. Establish a pest management policy stating that minimizing or eliminating pesticide use and choosing least-toxic alternatives is the goal.
2. Identify any pests and learn about their biology.
3. Determine at what point action needs to be taken (the action thresholds), that is, the point at which the pest problem is considered serious enough to require action.
4. Explore all nonchemical options, list each method's pros and cons; put in place programs that use the best of these options.
5. Keep written records and evaluate how the methods have worked; modify as necessary.

Pest Control in Schools

Many parents, teachers, and school administrators are examining the use of pesticides in schools because of concern that children and staff may be accidentally exposed to these chemicals. One study in New York State public schools made the following findings:

- Chemical pesticides that could cause short- or long-term health problems were used in most schools.
- Pesticides were used in many areas, including classrooms, offices, playgrounds, playing fields, bathrooms, basements, gyms, and especially in kitchens and cafeterias.
- The majority of schools did not post warnings for students or parents prior to pesticide use.

Many school districts have responded to concerns and adopted IPM programs aimed at eliminating the use of pesticides in schools. IPM is now being practiced by school districts across the United States, including parts of Maryland, California, Michigan, Illinois, and Pennsylvania.

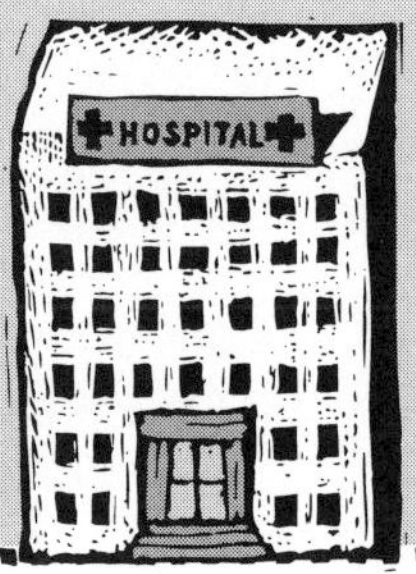

Section D

HUMAN HEALTH RISKS OF HAZARDOUS CHEMICALS

ISSUES AND INFORMATION

Hazardous chemicals can present countless risks to human health. Determining the risks is a complex task and government resources dedicated to this research are very limited. Scientists estimate that complete data on the hazards to human health and the environment is available for only two percent of the more than 70,000 chemicals in use today.

How chemicals enter our bodies is one factor in what effect they have on our health—that is, whether we breathe them in, swallow them, or come in contact with them on our skin or eyes. Certain substances are harmful if they enter our bodies through one route, but harmless through another route. Others are hazardous if they enter our bodies by any route.

Acute versus Chronic Effects

The amount of a chemical that enters the body is another factor in determining its health effect, including the amount of any one exposure as well as the number of exposures. Acute effects occur immediately after a single exposure to a hazardous chemical. Many chemicals cause acute effects, especially if they are not handled with care or figure in accidents involving children. In 1988, for example, about 9000 cases of poisoning related to household bleaches and 900 related to furniture polishes, oven cleaners, and fabric treatment products were reported. The most common acute effects of chemicals are eye, nose, or throat irritations; headaches; dizziness; and fatigue. Many of these effects disappear after exposure to the hazard is ended. However, the acute effect of contact with severely toxic chemicals can be much more serious and can even cause death.

Numerous chemicals have health effects if exposure, even in very small doses, is repeated over an extended period of time—days, weeks, months, or years. These effects are called chronic and are often serious and permanent. Many chemicals cause no health problem from one or two exposures but can result in severe illnesses such as cancer after chronic exposure. Breathing in the fumes of some solvents a couple of times, for example, may cause no harm, while breathing them in repeatedly can result in serious liver damage. Most chemicals have not been tested for their chronic effects.

Greater Risks for Children

While the danger of hazardous chemicals to the health of adults is great, the threat to children is even greater. Tests on hazardous chemicals often fail to take this fact into account. Children are more at risk for the following reasons:

- Children take in more food, air, and water in relation to their body weight than do adults. They also absorb into their bodies a greater percentage of substances that they eat or inhale.
- Children tend to breathe more through their mouths, allowing more particles to get into the lungs than when filtered through the nasal passages.
- Playing in dirt and on carpets, lawns, and other surfaces puts children in closer contact with contaminants. Young children also transfer many substances from their hands into their mouths.
- Children's immune systems are less developed and less capable of fighting off hazards.

- Children's brains and other parts of their nervous systems are still developing. Even small amounts of hazardous substances that attack the nervous system can damage their development.

Primary Health Effects

The major health risks of hazardous chemicals include cancer, gene mutation, birth defects, reproductive problems, nervous system problems, respiratory problems, skin problems, and damage to the liver or kidneys.

Cancer Many chemicals cause or promote the growth of cancer or cancerous tumors. Because cancer can develop slowly or be the result of small, repeated exposures to a chemical over time, and because many other factors come into play in human health, it is often difficult for researchers to establish for certain that a chemical causes cancer. Many chemicals, therefore, are identified as "possible" or "suspected" rather than as "known" carcinogens. Some medical scientists, however, estimate that 80 to 90 percent of human cancer is caused by contact with carcinogens in the environment. These carcinogens include pesticides; household chemicals such as glues and solvents (for example, those used in gasoline); toxic chemicals used in industry, in construction, in new furniture and carpets; alcohol, tobacco, drugs; chemical substances added to foods; air and water pollutants from industry and waste disposal.

Gene mutations Some chemicals cause mutations or alterations in genes. Gene mutation can lead to cancer, and the changed genes can pass on altered traits from parents to offspring. For example, one study found that men working in the metal industry were more than three times more likely to have children who developed brain cancer than men working in industries without exposure to hazardous metals.

Birth defects Birth defects include mental or physical problems that occur in the development of a baby before it is born. About 250,000 children are born with birth defects each year in the U.S., and the causes are certain for only about 40 percent of them. A number of chemicals—including lead, pesticides and prescription drugs—are known to cause birth defects; many others are suspected. Farm workers' children exposed to pesticides and the children of parents in industries manufacturing chemicals have particularly suffered from birth defects.

Reproductive problems Reproductive problems include the inability to have children or babies being born prematurely. Exposure to hazardous chemicals by either men or women who wish to be parents can cause reproductive problems. Chemicals that have been shown to harm reproductive organs are lead, certain prescription drugs, alcohol, tobacco, illegal drugs, and pesticides.

Nervous system problems Acute attack by hazardous chemicals on the nervous system can result in coma or death. Milder nervous system effects are headaches, nausea, and dizziness. Nervous system diseases including Parkinson's Disease and Alzheimer's Disease have been linked to exposure to hazardous chemicals. In addition, it has been discovered that chemicals attacking the nervous system can affect intelligence, language ability, moods and emotions, and attention. Studies have shown links between exposure to toxic chemicals and poor attention (or hyperactivity) in some children. Many different chemicals attack the nervous system, including pesticides, metals (such as lead and mercury), alcohols, rubber cements, and some aerosol sprays. The nervous system is also affected by most solvents, such as those used in some cleaning and dry-cleaning products, arts and crafts supplies (glues and quick-drying paints), and in gasoline.

Respiratory problems Chemicals that are contained in air as gases or small particles enter our lungs as we breath. They may cause harmful effects by lodging in our lungs or by absorption from the lungs into the bloodstream. Breathing in hazardous chemicals can cause respiratory problems, including coughing, shortness of breath, asthma, chronic bronchitis, pneumonia, emphysema, and lung cancer. Some of the chemicals known to cause problems in the respiratory system are room fresheners; deodorants; hair sprays; vapors from auto body shops; dry cleaning, paint, and gasoline vapors; automobile exhaust; and gases emitted from some new carpets and furniture.

Skin problems Contact with hazardous chemicals is a common cause of inflammations of the skin, including redness, swelling, and rashes. Various metals, dyes, rubber compounds, pesticides, cleaning products, waxes, and polishes are some of the chemicals that are hazardous to skin. In addition, some hazardous chemicals—including wood preservatives, other pesticides, and solvents in gasoline—can penetrate through the skin directly into the body.

Kidney and liver problems Kidney failure can result from exposure to hazardous chemicals such as solvents used in industry or heavy exposure to glues. Chronic kidney disease can develop from prolonged exposure over time to solvents such as those used in gasoline. Exposure to many solvents can also cause damage to the liver, ranging from mild damage to chronic disease and liver cancer. Some chemicals found in plastics, furniture and auto upholstery, packaging materials, and wall coverings have been linked to liver damage.

Section E

CHEMICALS AND THE FOOD CHAIN

Living organisms obtain energy by eating other organisms. The sequence of this transfer of energy from one organism to another is called a food chain or food cycle. For example, plants—which make their own food—are eaten by mice. Mice are eaten by snakes, and snakes are eaten by hawks. When the hawks die, they are eaten or are decomposed by bacteria, fungi, and other decomposers. This recycles the nutrients: hawks decompose to become food for the plants, and the cycle begins again. Organisms in food chains are further interconnected in food webs. When hazardous chemicals are introduced into the environment, they move from organism to organism through these interconnections.

Chemicals that last a long time are called "persistent." Persistent chemicals do not break down in the environment and may last for centuries, moving physically throughout the environment and cycling through food chains. Some pesticides that have been banned for decades because of their toxicity still persist in the environment today.

The effect of persistent chemicals is particularly devastating because they build up in the food chains in a process called "bioaccumulation." Because chemicals are stored in animal tissues and internal organs and because animals higher on the food chain eat more than animals lower down, chemicals increase in concentration—bioaccumulate—as they move up the food chain. For example, a water polluting chemical is absorbed in small quantity by aquatic organisms such as insects, microorganisms, and small fish. Small birds eat these organisms in great quantity and thus consume a greater amount of the chemicals (which are stored in the bodies of the first organisms). Mammals such as badgers or large birds such as eagles and hawks eat the smaller birds, taking in still larger concentrations of the contaminants. In these "top" organisms, the concentration of the chemical is high enough to cause diseases such as cancer and reproductive problems.

The pesticide DDT is probably the best-known example of bioaccumulation. DDT was developed and used extensively after World War II. It is an effective and very persistent pesticide. DDT, sprayed on plants to kill crop-destroying insects, was carried in irrigation and rainfall run-off into lakes and rivers where it contaminated aquatic organisms and plants. The plants and small organisms were eaten by fish. The fish were eaten by birds. In the birds, DDT causes the shells of their eggs to be so thin that

Bug-Be-Gone Moves Up the Food Chain

young cannot hatch. Because of this effect, DDT caused the near-extinction of many bird species including the bald eagle, the peregrine falcon, and the brown pelican. DDT is banned in the U.S., but it still persists in the environment and is still being used in many other countries.

This diagram shows how persistent pesticides (or other persistent chemicals) can accumulate and affect many organisms. Human beings, positioned at the top of many food chains, eating animals with high concentrations of chemicals, are particularly threatened by the hazards of chemicals used in the environment.

Persistent Pesticides Accumulate As They Move Up the Food Chain

Section F

ENVIRONMENTAL EFFECTS OF HAZARDOUS MATERIALS

In an attempt to improve people's standard of living, tens of thousands of chemical products have been developed. Sadly, much of the world's air, land, and water are now polluted with chemicals. Although early developers of chemicals were not as aware of environmental effects, we now know that a chemical introduced into one part of the environment can affect all the organisms and the entire balance of life connected to that environment. Pesticides that worked wonders at killing insects on farmlands have caused near extinction of ocean-going birds. Phosphate-containing detergents used in urban households have depleted fish populations in the Great Lakes. Many species of animals and plants have been harmed by environmental pollution from hazardous chemicals.

How Hazardous Chemicals Enter the Environment

One way hazardous chemicals are released into the environment is through direct application such as the use of pesticides, herbicides, and fertilizers on lawns and farms. These applications pollute the air and land and are carried by irrigation water into rivers, streams, and oceans.

Another source of hazardous chemicals in the environment is the disposal of hazardous wastes. Well over 300 million tons of hazardous waste are generated each year in the U.S. A great deal of hazardous waste comes from industry, including toxic solvents and heavy metal residues. A surprising amount, however, comes from households, including such chemicals as paints, solvents, cleaning supplies, and pesticides. These chemical wastes enter the environment when they leak into the soil and water supplies from garbage dumps, when they are released as air pollutants in the burning of garbage, and when they are dumped carelessly into household drains, storm drains, or natural waterways. Medical wastes have been found washing up on the shores in New Jersey and New York, for example.

Accidental release is a third manner in which hazardous chemicals get into the environment. Hazardous substances may leak out of buried storage containers and enter the soil and underground water. Also, industrial accidents can happen, such as the release of toxic gas from a manufacturing plant in Bhopal, India, which killed more than 2000 people within a few hours.

How Hazardous Chemicals Affect the Environment

Chemical pollutants know no boundaries—they can travel hundreds of miles in the air, through the soil, and in the surface and underground waters. They interfere with biological processes in both plants and animals and, just as they affect human health, they can cause damage to the health of other living organisms.

Air Pollution Persistent chemicals can be carried in the air as gases or small particles. Traveling in the air, they are carried into our bodies, foods, water, and wildlife. Following are some examples of air pollution from hazardous chemicals and their environmental effects:

- Pesticides, particularly those that are sprayed, travel great distances. These chemicals can cause serious short- or long-term damage to many plants and animals in addition to those intended to be harmed. The Environmental

Protection Agency has reported finding pesticides in the Great Lakes that were carried by air currents from as far away as Central America.

- The burning of waste in incinerators produces air pollution containing dangerous vaporized heavy metals—including lead, cadmium, and mercury—and other poisonous gases. These poisons can impair the health of organisms in the environment as well as threaten human health.
- Some gaseous chemicals used as aerosol sprays, solvents in industry, and in air conditioners may be causing the destruction of the earth's ozone layer. This in turn is leading to greater exposure to ultraviolet rays from the sun, which causes crop failures, skin cancer and other diseases, and may be affecting global weather patterns.
- Chemical industries sometimes release hazardous by-products into the air as gases or small particles. These may cause illness when breathed in by animals or may enter the food chain when they fall on land or water.
- Some building materials—including plywood, particleboard, foam insulation, and floor tile—and some new furniture and household carpets and textiles contain chemicals that give off hazardous fumes. In homes or buildings, particularly if they are well-sealed from outside air, the fumes from these materials can be harmful to human health and may even cause cancer.
- Many household products such as paints, thinners, solvents, cleaning chemicals, glues, and hobby supplies release fumes that are hazardous—especially in unventilated areas—to humans and pets.

Land Pollution Hazardous chemicals pollute the land from sources such as household waste, industrial waste, and pesticide or herbicide application. On the land, these chemicals can destroy plant life, harm animals that come in contact with them, or seep into underground water or surface waters, harming aquatic life and all organisms in the food chain. It is sometimes hard to imagine that our everyday lives pollute the land, but, on average, each person in the U.S. generates about 160 pounds of household hazardous waste each year. This waste comes from discarding such materials as toilet bowl cleaners, window cleaners, furniture polishes, disinfectants, paints, varnishes, motor oil, and pesticides. Thrown away in the trash, these chemicals can wind up leaking from garbage dumps into the soil and water supplies or polluting the air as particles from garbage incineration.

In the past, many industries simply dumped their waste materials on the land. While regulations govern the disposal of waste today, hazardous chemicals from industrial waste can pollute the land if they leak out of underground containers or garbage dumps, or if industries irresponsibly dispose of waste.

Pesticides are among the most hazardous sources of land pollution. Their harmful effects can be very widespread and long-term. (See Issues and Information sections C and E.)

Water Pollution Water pollution is one of our most serious environmental problems, and hazardous chemicals and metals are among the principle water contaminants. The U.S. government has identified over 700 pollutants in drinking water, some of which have been found in every state in the nation.

Sources of hazardous chemicals in water include

- industrial waste discharged directly into surface waters or sewers
- improper disposal of chemicals such as cleaning products, paint removers, art supplies, solvents, and motor oil
- run off or seepage of pesticides and fertilizers from farmland or yards into surface waterways or underground water
- seepage from garbage dumps or industrial or illegal disposal sites

Hazardous chemicals in common use, if poured down a drain or toilet or put into the trash, can leak from the garbage dump or septic tank or pass untreated through the sewage system. Thus they make their way into the water supplies, where they cause harm to human health and contaminate ponds, streams, wetlands, and oceans. These polluted waters can destroy the balances of aquatic plant and animal life.

In the 1970s, for example, entire natural populations of plants, fish, and other aquatic creatures were destroyed in the Great Lakes by the huge amounts of phosphates (about half of which were from household laundry detergents) contaminating the waters. These chemicals over-fertilize algae in the lakes, which grow to the point that they block oxygen and sunlight from other species of plants and animals, disrupting the entire natural system.

The many harmful results of water pollution include damage to life in the ocean. Fish have been found with tumors and lesions, crustaceans have been found with chemical burn holes, and shellfish populations have been destroyed. By 1990 in the continental U.S., about one-third of the coastal waters previously open to shellfish harvesting were closed because of pollution and habitat destruction.

The effects of drinking polluted water on human health are numerous. They include cancer; nervous system disorders; intestinal problems; bone damage; and damage to the heart, liver, and kidneys (see Issues and Information section D).

Section G

LISTS OF SELECTED HAZARDOUS CHEMICALS

ISSUES AND INFORMATION

No one knows exactly how many hazardous chemicals exist in the world today or what they all are. We do know that there are thousands of chemicals that can cause severe harm to human health and to the environment. Many have been tested and classified as hazardous (or toxic), many have shown harmful effects but have not been tested enough to prove their hazardous characteristics, and many have not been tested at all. With that in mind, here are a couple of lists to give you an idea of some hazardous substances to be aware of.

This first list is of some hazardous products commonly used in schools and homes, classified by their use. Remember, many different products are often available for the same use; some of the alternatives may be much less harmful than others.

Use	Hazardous Products
Automotive	antifreeze batteries brake fluid car wax and cleaners gasoline and diesel fuel lead-acid batteries motor oil transmission fluid windshield wiper fluid
Building Maintenance and Repair	adhesives caulk glue paint paint primer paint stripper paint thinner and paint brush cleaner stain varnish
Pest Control	disinfectant flea collars insecticide and insect repellent mothballs pet spray and dip rat and mouse poison weed killer wood preservative
Cleaning	ammonia bleach degreasing products deodorizer drain opener floor care products furniture polish furniture wax metal polish oven cleaner spot remover toilet bowl cleaner tub and tile cleaner
Lawn and Garden Care	fertilizer fungicide herbicide insecticide pesticide
Other	arts and crafts supplies cosmetics household batteries lighter fluid particle board and plywood photographic chemicals pool chemicals prescription medicines shoe polish

To give you an introduction to a few specific chemicals and their effects, here are some of the chemicals that are identified by the U.S. Department of Health and Human Services' Agency for Toxic Substances and Disease Registry as being among the "Top 20 Hazardous Substances" for 1995.

Chemical and Its Sources	Human Health Effects
Lead Storage batteries in waste dumps; paint chips eaten by children; gasoline additives emitting fumes; metal pipes and solder in plumbing.	Reproductive problems; decreased intelligence, slow growth, hearing problems in children; possible cancer; brain and kidney damage.
Arsenic Pesticides contaminating soil and water; leakage from waste sites and metal manufacturing plants; natural deposits contaminating water.	Large doses cause death; lower exposures cause skin abnormalities; digestive problems; circulatory system problems; liver and kidney damage; nerve function disorder; increased risk of skin cancer; and cancer of liver, bladder, kidney, and lungs.
Vinyl Chloride Plastics and vinyl products including PVC pipes, cable coatings, packaging materials, furniture and auto upholstery, wall coverings, etc. Enters environment when released into air and discharged into wastewater from plastics industry, or evaporates or leaks from chemical waste storage sites. Air inside new cars may have high levels before aired out.	A known carcinogen. Illnesses range from dizziness, lack of muscle coordination, headache, and cancer to death. Chronic exposure causes "vinyl chloride disease," including severe damage to liver, lung damage, poor circulation, thickening of skin, and changes in blood.
Benzene Industrial solvent for making other chemicals, contained in some plastics, detergents, pesticides, glues, cleaning products, paint strippers, art supplies, tobacco, and gasoline; exposure usually through breathing vapors.	High doses cause dizziness, drowsiness, headaches, and may cause death. Low chronic doses may cause cancer; harms immune system causes gene mutations, possible birth defects.
PCBs (polychlorinated biphenyls) Includes over 200 chemicals. Used widely in electrical equipment; very persistent, bioaccumulates; enter air, soil, and water from waste sites, industrial dumping, from old fluorescent lighting fixtures and electrical devices; humans contaminated from eating contaminated fish and in buildings that contain PCB-using devices.	Probable human carcinogen, suspected of causing human reproductive and developmental problems; causes liver damage, skin irritations, reproductive and developmental problems, and cancer in animals; the manufacture of PCBs halted in 1977 in the U.S. due to health hazards, but persists in the environment.

Chemical and Its Sources	Human Health Effects
Chloroform Air conditioners, paper mills, chemical manufacturing, wastewater from sewage treatment and chlorinated drinking water (chlorinating is for disinfecting purposes); used in pesticides, dyes, fire-extinguishers, dry cleaning, swimming pools. Solvent in paint supplies (lacquers, greases, waxes, adhesives, oils). Travels long distances in air. Increased exposure is a danger for people working in paper, pesticide, food-processing, and chemical plants and in paint stores.	Damage to central nervous system, liver, and kidneys. Short-term exposure causes tiredness, dizziness, headache. Suspected carcinogen; causes liver and kidney cancer in animals.
DDT, DDE, and DDD Pesticides. Banned in U.S. since 1970s due to health effects. Persist for a long time bound to soils. Found today in foods, including root and leafy vegetables, meat, fish, and poultry. Still used outside the U.S.; imported foods may cause exposure U.S. regulations have tried to avoid.	Once in body, stored readily in fatty tissue. Affects nervous system and possibly affects liver and reproduction. Listed as suspected human carcinogen.

Section H

SAFETY PRACTICES FOR USING CHEMICAL PRODUCTS

The best way to deal with the safety problems of hazardous chemicals is to use nonhazardous alternatives instead. However, it may be possible to only change our choices and behaviors gradually. Here are some tips to reduce the use of hazardous chemicals and to help reduce their impact if they are used.

Product Choices

- Compare products and choose the least-harmful ones.
- Avoid powders or aerosols that create air pollutants.
- Avoid products containing solvents that emit vapors (some glues, paint thinners, arts and crafts supplies).
- Buy organic produce that is grown without the use of pesticides.

Precautions

Chemicals are dangerous, and you should seek advice on the use of specific substances at your local department of health, the Environmental Protection Agency, the local agricultural extension office, or your state agency for pesticides and other toxic chemicals. If you do use hazardous chemicals, practicing these safety measures will help reduce their harmful effects.

- Read product labels carefully to know the hazards of the material. If it is available, read the manufacturer's material safety data sheet (MSDS) for more complete information.
- Know what to do if an accident occurs and follow instructions thoroughly; keep first-aid, fire, and safety equipment nearby.
- Use only the amount of the substance that the instructions suggest. Never think that twice the amount is twice as good.
- Wear protective equipment (masks, goggles, gloves). If pesticides call for "restricted use" or for using protective equipment, though, it is best to leave the application to a certified expert.
- Never mix chemical products unless you know it is safe to do so.
- Use chemicals in well-ventilated areas. An open window rarely provides sufficient ventilation. If your use is outdoors, close all nearby windows.
- Keep hazardous substances well out of the reach of children or pets.
- Avoid wearing contact lenses when using chemicals that emit vapors (the contact lenses absorb the vapors and cause eye damage).
- Thoroughly wash up and clean up the work area after using hazardous materials; be sure to close all containers.
- Store all products in their original containers with their original labels.
- Store materials in cool, well-ventilated areas, away from children, pets, and food.

Disposal of Hazardous Waste

- Hazardous waste should be taken to a municipal drop-off center or to a licensed hazardous waste disposal company. Find out about hazardous waste collection in your community (call your city or county health and safety departments or your local solid waste management company).
- Never pour hazardous chemicals down inside or outside drains or toilets. Never throw them away in the trash.

GLOSSARY

carcinogen A cancer-causing substance. Between 150 and 200 chemicals appear on various government agency lists as having sufficient or limited evidence that they are carcinogenic.

chemical A substance formed by altering the molecular structure of a natural material, or the combining of two or more elements to form a new substance.

corrosive Capable of eating or wearing away gradually, by rusting or by the action of chemicals.

flammability The capability of burning or igniting.

genetic mutation A sudden, unpredictable change in heredity resulting in an individual unlike the parents; caused by a change in a gene or chromosome.

hazardous Having the ability to cause harm; flammable, explosive, irritating or damaging to skin or lungs (such as are strong acidic or alkaline substances), or inducing allergic reactions in the immune system. Many hazardous materials require special handling and disposal.

herbicide A substance or preparation for killing plants, especially those considered to be weeds.

inert ingredients Any ingredient other than an active ingredient. For example, almost all pesticide products contain some inert ingredients; they are used as solvents, surfactants, to dilute the product, and so on.

insecticide The most common pesticides are the insecticides, which are chemical agents applied to the environment to kill insects.

integrated pest management (IPM) A strategy for controlling pests using methods that emphasize source reduction and prevention, along with minimal use of low toxicity pesticides.

organic Refers to plants and animals grown without drugs or synthetic fertilizers or pesticides; carbon-based compounds, such as benzene and turpentine.

pesticide Any chemical preparation used to control populations of organisms, including plants and animals, thought to be damaging to that organism. Most pesticides are not highly selective in their action and are introduced to control plant diseases, increase agricultural productivity, control urban pests, and preserve building materials.

petrochemical Any substance derived from crude oil.

reactive Tending to react with other substances or chemicals.

solvent The dissolving medium or liquid portion of a solution. (Water is frequently referred to as the universal solvent.) The term solvent also applies to organic materials such as benzene, acetone, or gasoline, used to clean oils or grease from machinery, fabrics, or other surfaces. Many organic solvents are flammable and/or toxic.

surfactant A surface-active agent that decreases the surface tension of water, making the water "wetter" and better able to penetrate fabrics or other materials in order to lift dirt and grease.

synthetic chemical A chemical formed by combining single elements into a more complex structure or a unique material not found in nature. Many synthetic ingredients are at least in part derived from petroleum products.

toxic Tending to cause acute or chronic damage to biological tissue following physical contact such as ingestion, inhalation, absorption through the skin, and so on; poisonous.

toxicity The relative or specific degree of being toxic.

Teacher Resources

ORGANIZATIONS

Organizations Providing Educational Materials and Information

Art & Creative Materials Institute, Inc., The (ACMI)

100 Boylston Street, Suite 1050
Boston, MA 02116
(617) 426 6400
Fax: (617) 426-6639

The ACMI sponsors a certification program for children's and adult's art materials, certifying those materials are nontoxic and meet voluntary standards of quality and performance or are affixed with health warning labels where appropriate. They provide a list of ACMI-certified products, a list of current members, and a booklet, *What You Need to Know About the Safety of Art & Craft Materials,* free of charge.

Alliance to End Childhood Lead Poisoning

227 Massachusetts Ave. Suite 200 NE
Washington, DC 20002
(202) 543-1147
Fax (202) 543-4466
E-mail: <aeclp@permanet.org>

A national nonprofit public interest organization created to launch a comprehensive attack on the epidemic of childhood lead poisoning. Products & Services: The Alliance's education efforts inform health professionals, the general public, and policy makers of lead hazards and the need for prevention. *Childhood Lead Poisoning: Blueprint for Prevention; National Benchmark Standards for Lead-Safe Housing and International Action Plan for Preventing Lead Poisoning.*

Bio-Integral Resource Center

PO Box 7414
Berkeley, CA 94707
(510) 524-2567
Fax: (510) 524-1758
E-mail: <birc@igc.apc.org>

BIRC is a nonprofit organization formed to provide practical information on least-toxic methods for managing pests, based on the principles of Integrated Pest Management, or IPM.

Products & Services: Publications on Integrated Pest Management, pests of the home and garden, pests of people and pets, agricultural and community-wide pests, and biological control and least-toxic pesticides.

Center for Safety in the Arts

c/o NYFA
155 Avenue of the Americas, 14th Floor
New York, NY 10013
E-mail: <csa@artswire.org>
WWW:
<http://gopher.tmn.com:70/1/Artswire/csa>

The center is a national resource center for research and education on hazards in the visual arts, performing arts, arts programs in educational facilities and museums.

Products & Services: Books, pamphlets, and numerous articles, such as "Children's Art Supplies Can Be Toxic" and quarterly newsletter, *Art Hazards News.*

Environmental and Occupational Health Sciences Institute

681 Frelinghuysen Road
P.O. Box 1179
Piscataway, New Jersey 08855-1179
(908) 445-0110
Fax: (908) 445-0122

The EOHSI houses a select group of scientists, physicians, educators, and policy researchers focused on the serious health effects of environmental pollutants and hazards.

Products & Services: The efforts of the Resource Center of EOHSI are geared toward grades K-12, and feature: a free summer institute for educators, curricular materials including video and computer software components, INFOsheets (informational brochures which provide an overview of specific environmental and occupational health topics), teacher newsletter, interactive television training, and more.

Ethical Science Education Coalition

167 Milk Street #423
Boston, MA 02109-4315
(617) 367-9143
(860) 878-1808
Fax: (617) 523-7925
E-mail: <info@ma.neavs.com>

A national organization of educators, students, and concerned citizens who want to ensure that a student's right to a quality education need not be compromised by their ethical belief regarding the use of animals in the classroom.

Products & Services: Assists schools in formulating and implementing dissection policies and disseminates information regarding the alternative non-animal teaching tools available. Provides the following: *Frog Fact Sheet; Turtle Fact Sheet; Choice Dissection Policy Sheet, Beyond Dissection-Innovative Teaching Tools for Biology Education.*

Household Hazardous Waste Project

1031 East Battlefield, Suite 214
Springfield, MO 65807
(417) 889-5000
Fax: (417) 889-5012
E-mail: <househol@ext.missouri.edu>

The Household Hazardous Waste Project is a program of the University of Missouri Extension System in cooperation with the Environmental Improvement and Energy Resource Authority. HHWP encourages household hazardous product identification, safe use, storage, proper disposal, and the selection of safer alternatives. HHWP's purpose is to protect the quality of health and environment through increasing people's knowledge and ability to make informed decisions about hazardous products and unregulated wastes.

Products & Services: HHWP provides educational materials, training, consulting, and a referral-and-information service for concerns regarding household hazardous products. *Guide to Hazardous Products Around the Home* is a handbook that explains product ingredients, health and safety issues, disposal, recycling outlets, and safer product alternatives; *Guide Sheet Series* are guide sheets that answer commonly asked questions about household hazardous products and provide guidelines for safe use, storage, and disposal.

Human Ecology Action League, Inc. (H.E.A.L.)

P.O. Box 29629
Atlanta, GA 30359-1126
(404) 248-1898
Fax: (404) 248-0162
E-mail: <HEALNatal.aol.com>
WWW: <http://home.aol.com/HEALNatnl>

HEAL is a nonprofit national environmental health organization serving people with multiple chemical sensitivities (MCS) and others concerned about the effects of chemicals on human health. Their mission is to educate and inform about exposure to the modern chemicals increasingly found in perfumes, fabric softeners, detergents, paints, pesticides, inks, fabric, tobacco smoke, building materials, auto exhaust, and paper.

Products & Services: *The Human Ecologist* offers suggestions for coping with environmental stresses; sources for food, clothing, and household supplies; news of advances in medicine, science, and government regulations; ideas for working effectively with agencies and health-care professionals; brochures/flyers *(Chemicals Can Affect Your Health)* explaining chemical sensitivities.

Mothers and Others for a Livable Planet

40 West 20th Street
New York, NY 10011
(212) 242-0010
(212) 888-INFO
Fax: (212) 242-0545
E-mail: <mothers@igc.apc.org>

A national nonprofit consumer education group, co-founded by Merle Streep and dedicated to bringing everyday environmentalism into homes and communities by providing information that is accessible and practical.

Products & Services: Informs consumers that alternatives do exist to such environmental dilemmas as pesticide-laden foods, and mobilizes consumers to demand healthy choices in their supermarkets. Mothers & Others is a membership organization. Membership ($25) includes a subscription to a newsletter, *The Green Guide;* free fact sheets and discounts on publications. Fact sheets cover issues such as synthetic bovine growth hormone, least-toxic pest control, reasons to buy organic, and books such as *The Way We Grow: Good-Sense Solutions for Protecting our Families* are also available to nonmembers.

National Anti-Vivisection Society

53 W. Jackson Blvd., Suite 1552
Chicago, IL 60604
(800) 888-6287
(800) 922-FROG Hotline

NAVS believes that how you learn is as important as what you learn. Focus of the hotline is to encourage students' respect for animals; foster understanding and tolerance for students who object to dissection; and promote a gentle, nonviolent educational experience.

Products & Services: They publish several booklets. *Objecting to Dissection* provides students with the tools to object effectively to dissection and numerous resources (models, videos, computer programs) for alternatives to dissection.

National Coalition Against the Misuse of Pesticides

701 E Street SE, Suite 200
Washington, DC 20003
(202) 543-5450
Fax: (202) 543-4791
E-mail: <ncamp@igc.apc.org>

NCAMP was established by a broad coalition of health, environmental, labor, farm, consumer, and church groups, as well as individuals, who share common concerns about the potential hazards associated with pesticides. The coalition seeks to focus public attention on the very serious pesticide poisoning problem and promotes reduced pesticide exposure through alternative pest management strategies that use few or no toxic chemicals. The coalition also advocates for public policies that better protect the public from pesticide exposure, and it assists community organizations with information on alternative methods of pest control.

Products & Services: *Safety at Home: A Guide to the Hazards of Lawn and Garden Pesticides and Safer Ways to Manage Pests;* packets on *Fertilizers, Pesticides on Food, Mercury in Paint, School IPM; Pesticides in Schools.*

National Pesticide Telecommunications Network

(800) 858-7378, M–F, 8:30 AM–4:30 PM
Central Time

Hotline sponsored by Oregon State University. Provides scientific and factual information about pesticides; safety, health effects, and environmental effects. Will provide referrals for lab analysis in the case of human and animal poisonings and information on lawn care, pest control, pesticides, and soil testing.

National Soy Ink Information Center

1025 Ashworth Road, No. 310
West Des Moines, IA 50265-3542
(800) 747-4275
Fax: (515) 223-4331

Provides information on soy ink, the SoySeal, or a list of soy ink manufacturers.

People for the Ethical Treatment of Animals (PETA)

501 Front Street
Norfolk, VA 23510
(757) 622-PETA
Fax: (757) 622-0457
WWW: <http://envirolink.org/arrs/peta>

PETA is an international, nonprofit organization dedicated to exposing and eliminating animal abuse. PETA uses public education, litigation, research and investigation, media campaigns, and grassroots organizing to accomplish its goal of establishing and defending the rights of animals.

Products & Services: Offers educational brochures and fact sheets.

Rachel Carson Council

8940 Jones Mill Road
Chevy Chase, MD 20815
(301) 652-1877
Fax: (301) 951-7179
E-mail: <rccouncil@aol.com>

Resource center that offers educational publication on the use of and alternatives to pesticides.

Products & Services: *Basic Guide to Pesticides; Alternative Pest Control for Lawns and Gardens; Least-Toxic Ant Control; Least-Toxic Cockroach Control; Pesticides and Lawns* (Locate by table the active ingredients in 72 commonly used pesticides. Learn the characteristics and hazards of these lawn and garden chemicals). *The Other Road to Flea Control* (user friendly guide to IPM in pet care); *Toxic Playgrounds* (discussion of the use of wood treated with Chromated Copper Arsenate [CCA] to build playground equipment).

GOVERNMENT AGENCIES

Environmental Protection Agency Programs

Indoor Air Quality Hotline

(800) 438-4318, M–F, 9 AM–5 PM EST

Will answer questions on IAQ, problems and remedies. Also offers an extensive publication list, including pamphlets on secondhand smoke, asbestos, formaldehyde, radon, how to maintain HVAC systems, hazards in the science room, and many more.

National Lead Information Center

(800) LEAD FYI
National Safety Council
1019 19th Street NW, Suite 401
Washington, DC 20036-5105

NLIC is operated by the National Safety Council, with funding from the EPA, Centers for Disease Control, and Department of Housing and Urban Development. The hotline takes names and addresses for mailing information on the hazards of lead, testing for lead, how to avoid lead poisoning, dealing with lead poisoning, as well as state contacts for additional information.

National Radon Helpline

(800) 55-Radon
Information specialists answer questions on various radon issues; disseminate literature on radon testing, radon mitigation, and reducing radon in new construction; and can refer people to other agencies as appropriate.

National Radon Hotline

(800) SOS-Radon

Operated by the National Safety Council through a grant from the EPA. The hotline disseminates general information on radon, and a coupon for a low-cost radon test kit.

Office of Pesticide Programs

Waterside Mall
401 M Street, SW, MS 7506C
Washington DC 20460
(703) 305-7090
Responsible for regulating the use of pesticides and registering pesticides in the United States.

Products & Services: Provides a number of brochures and fact sheets on the safe use, storage, and disposal of pesticides. It provides a care packet that includes most of their publications, such as *IPM in Schools* and *Healthy Lawn, Healthy Environment* and the *Protecting Endangered Species from Pesticides* poster.

Radon Test Measurements in Schools

Public Information Center (PIC)
401 M Street SW, #3404
Washington, DC 20460
(202) 260-7751
Fax: (202) 260-6257
E-mail: public-access@epamail.epa.gov

Call or write to obtain a 44-page booklet, *Radon Measurement in Schools.* Contact your regional EPA office to obtain list of recommended companies to measure the school for radon.

U.S. Consumer Product Safety Commission

Washington, DC 20207

If you find lead in your home, send a postcard to the address above and ask for the free pamphlet *Protect Your Family From Lead in Your Home.*

BOOKS AND PAMPHLETS

A Building Revolution: How Ecology and Health Concerns Are Transforming Construction

David Malin Roodman and Nicholas Lenssen
Worldwatch Paper 124, March 1995
World Watch Institute
1776 Mass. Avenue NW
Washington, DC 20036
(202) 452-1999
Fax: (202) 296-7365
E-mail: <worldwatch@worldwatch.org>
WWW: <http://www/worldwatch.org>

Informational document on how a variety of traditional design solutions and advanced technologies, from earthen materials to efficient lights, can cost effectively eliminate almost all of the damage done by modern buildings while preserving the comfort and amenities that people expect. Also explains how structures that pose fewer environmental and health problems are better places to work and live. 67 pages.
$5.00

A Consumer's Dictionary of Cosmetic Ingredients

Ruth Winter
New York: Crown Publishers, 1994

A dictionary-type listing of the harmful and desirable ingredients found in cosmetics. 410 pages.
$14

A Consumer's Dictionary of Household, Yard, and Office Chemicals

Ruth Winter
New York: Crown Publishers, Inc., 1992

A dictionary-type listing of chemicals and their derivation, use, benefits, hazards, and substitutes. This book presents the extent of the problem, the government and industry regulation—or lack of regulation—of the chemicals, and what consumers can do to protect themselves and their families.
$12.00

Artist Beware

Michael McCann
New York: Lyons & Burford Publishers, 1992

The first part of the book is a general introduction to the chemical and physical hazards commonly found in art and craft materials and processes, and suitable precautions. The second part of the book discusses the hazards of particular techniques, the process and the materials used, and precautions to take for protection. 564 pages.
$29.95

Clean & Green

Annie Berthold-Bond
Ceres Press, 1994
P.O. Box 87
Woodstock, NY 12498
(914) 679-5576

Filled with new and traditional ideas for living free of hazardous household products. This extensive guide to nontoxic and environmentally safe housekeeping provides 485 ways to clean, polish, disinfect, deodorize, launder, remove stains, and even wax your car without harming yourself or the environment. 162 pages.
$8.95

Common Sense Pest Control: Least-toxic solutions for your home, garden, pets and community

William Olkowski, Sheila Daar, Helga Olkowski
Taunton Press, 1991
63 South Main Street
P.O. Box 5506
Newtown, CT 06470

A complete reference guide for pest management programs for indoor and outdoor landscapes, greenhouses, buildings, animals, the human body, and more. 715 pages.
$39.95 plus shipping

Environmental Hazards in Your School: A Resource Handbook

TSVA Assistance Information Service (TAIS) 7408
401 M Street, SW 3404
Washington, DC 20460
(202) 554-1404
Fax: (202) 554-5603
E-mail: <tsca-hotline@epamail.epa.gov>

An overview of asbestos, indoor air quality, radon, lead, and other environmental concerns, with resources for taking action if your school faces a problem. 60 pages.

Health Hazards Manual for Artists

Michael McCann, Ph.D., C.I.H.
New York: Lyons & Burford Publishers
A guide for teachers and artists divided into three parts: how art materials affect you, hazards of various media, and safety in the studio.
$11.95

Healthy School Handbook, The

Norma L. Miller, Ed.D, Editor
West Haven, CT: National Education Association Professional Library Publication, 1995

Twenty-five professionals collaborated to produce this handbook for identifying and conquering the sick building syndrome and other environmental hazards in and around the school. 446 pages.
$24.95

Indoor Air Quality Tools for Schools

Superintendent of Documents
P.O. Box 371954
Pittsburgh, PA 15250-7954
(202) 512-1800
Fax: (202) 512-2250

This easy-to-use action kit shows how to carry out a practical indoor air action plan at little or no cost, using common-sense activities and in-house staff. The kit includes checklists with instructions for all school employees, background information, sample memos and policies, a recommended IAQ Management Plan, and an Indoor Air Quality Problem Solving Wheel.
$22

Inside Story: A Guide to Indoor Air Quality, The

EPA's IAQ Information Clearinghouse
(800) 438-4318 9 AM–5 PM EST
To order:
Superintendent of Documents
P.O. Box 371954
Pittsburgh, PA 15250-7954

A 35-page booklet that offers an introduction to IAQ problems. Although it centers around the home, the sources and solutions are the same for a school.
$44 for 25 copies

Least Toxic Home Pest Control

Dan Stein
Eugene, OR: Hulogosi Communications, Inc., 1991

A humorous and practical guide to solving most common household/school pest problems with an absolute minimum amount of insecticides. 86 pages.
$8.95

Non-Toxic Home & Office, The

Debra Lynn Dadd
Los Angeles: Jeremy P. Tarcher, Inc., 1992

This is subtitled *Protecting Yourself and Your Family from Everyday Toxics and Health Hazards.* The book covers a wide range of products and substances that are potentially harmful and offers ways to eliminate them from your life and environment.
$10.95

Pest Control in the School Environment: Adopting Integrated Pest Management

U.S. Environmental Protection Agency, August 1993
Public Information Center 3406
401 M Street SW
Washington, DC 20460
(202) 260-2080
Fax: (202) 260-6257
E-mail: <public-access@epamail.epa.gov>
WWW: <http://www.epa.gov> or: <http://www.epa.gov/docs/oar-appd.html>

This booklet provides students with a general understanding of IPM principles so that they can make informed decisions about pest control in their schools. 43 pages.

Poisoning Our Children, Surviving in a Toxic World

Nancy Sokol Green
The Noble Press, Inc., 1991

Outlines ways in which all of us can eliminate exposure to dangerous chemicals. Included is information on the pesticide residues in food; the formaldehyde in furniture, carpet, and clothing; the untested chemicals in household cleaning products—how and why they can make us sick, and what to do about it. 263 pages. $12.95

Radon Test Measurements in Schools

Indoor Air Quality Information Clearinghouse
P.O. Box 37133
Washington, DC 20013-7133
(800) 438-4318
(202) 484-1307
Fax: (202) 484-1510

Call or write to obtain the booklet, *Radon Measurement in Schools* EPA-402-R-92-014. Quantities are limited so there is a limit of one document per school. Contact your regional EPA office to obtain a list of recommended companies to measure the school for radon. 44 pages.

PRODUCTS AND SERVICES

AFM (American Formulating and Manufacturing)

350 W. Ash Street, Suite 700
San Diego, CA 92101
(619) 239-0321
Fax: (619) 239-0565

A catalog of safe, nontoxic products for the home and industrial environment. Products include paints, cleaners, caulking, adhesives, stains, and sealers.

Allergy Resources

P.O. Box 444
Guffey, CO 80820
(800) USE FLAX
Fax: (719) 689-2969

A free mail order catalog designed to meet the needs of allergenic and chemically sensitive individuals. Includes air purifiers, cosmetics, cleaning products, bedding.

American Environmental Health Foundation, Inc.

8345 Walnut Hill Lane
Suite 225
Dallas, TX 75231
(214) 361-9515
(800) 428-2343
Fax: (214) 361-2534
WWW: <http://www.aehf.com>

AEHF provides publications covering environmentally triggered diseases; products designed especially for chemically sensitive individuals (i.e., ceramic and stoneware oxygen masks, vitamin and mineral supplements, water and air filters, syringes, radon and formaldehyde test kits); materials for the education and training of students.

Auro Organics Natural Plant Chemistry

Sinan Company, Natural Building Materials
P.O. Box 857
Davis, CA 95617-0857
(916) 753-3104
E-mail <sinan@dcn.davis.ca.us>
WWW: <http://www.dcn.davis.ca.us/go/sinan>

Offers a catalog including nontoxic, plant-derived, petroleum-free wood protectors, stains, paints, waxes, linseed oil, organic soaps, beeswax floor care, natural resin lacquers, glues, natural cleansers and polishes, and furniture products.

BCD International

A Division of RPS International, Inc.
1935 Camino Vida Roble
Carlsbad, CA 92008
(800) 772-9339
Fax: (619) 431-9263

Provides safe, effective alternative cleaners for industrial workers, mechanics, and do-it-yourselvers. Their products meet stringent international ecological testing requirements. Products offered include: Ultra-concentrated Multi-Purpose Cleaner & Degreaser, Graffiti Remover & Multi-purpose Solvent, and Graffiti Barrier.

Carolina Environment

P.O. Box 26661
Charlotte, NC 28221
(800) 448-LEAD

Call to locate store closest to you selling KNOW LEAD lead detection kits. This kit aids in eliminating long-term effects of lead in a child, which can be severe and sometimes devastating (learning disabilities, hyperactivity, decreased growth, kidney and heart disease, and even brain damage) by detecting lead on any surface.

Products & Services: Various brochures on lead and poisoning, *The Shopper's Guide to Low-Lead China* , released by the Environmental Defense Fund and the Attorney General of the State of California, gives the names of individual patterns of fine and everyday china that meet California's stringent standards on lead exposure. The *Know Lead* brochure describes symptoms, effects, and sources of lead poisoning.

Eco Design Co./The Natural Choice

1365 Rufina Circle
Santa Fe, NM 87505-2965
(505) 438-3448 or (505) 438-0199

Catalog of natural paints (interior and exterior), stains, craft paints, and so on. They also offer a catalog of healthy home products.

Environmental Education and Health Services, Inc.

P.O. Box 92004
Austin, TX 78709-2004
(512) 288-2369
Fax: (512) 288-9538

EEHS was established by Mary Oetzel to provide educational and consulting services to individuals, schools, businesses, and municipalities concerned about indoor air quality. She will assist schools in identifying and remediating air quality problems in existing school buildings, review and make recommendations on architectural specifications for building or remodeling, and conduct presentations to generate awareness regarding indoor air quality issues.

Lead√Check Swabs

HybriVet Systems, Inc.
P.O. Box 1210
Framingham, MA 01701
(800) 262-LEAD
(508) 651-7881
Fax: (508) 651-8837

Lead√Check Swabs are instant screen tests for the detection of lead on solid surfaces, which include: paint, soil, pipe solder, dust, mini-blinds, and porcelain bathtubs. Call to find a local source for the swabs.

The Living Source

P.O. Box 20155
Waco, TX 76702
(817) 776-4878
(800) 662-8787
Fax: (817) 776-9329

Provide a catalog of products for the environmentally aware and chemically sensitive, ranging from personal care to books, nutritional supplements, building products and equipment, to pet supplies, cleaners, and more.

Naturlich Flooring

P.O. Box 1677
Sebastopol, CA 95473-1677
(707) 824-0914
Fax: (800) 329-9398

Natural floor covering; wool carpeting, true linoleum, cork, sisals and grasses, area rugs and accessories.

N.E.E.D.S.

National Ecological and Environmental Delivery System
527 Charles Avenue 12A
Syracuse, New York 13209
(800) 634-1380
Fax: (315) 488-6336

Discount supplier of products for those with allergies, asthma, and multiple chemical sensitivities. It also supplies alternative health products, including nutritional supplements; air and water purification systems.

VRP Marketing, Inc.

SoyClean Information
9135 Barton Street
Overland Park, KS 66214
(913)599-0800
Fax: (913) 599-2923

Offers industrial cleaning products that are soybean based. Cleaners include SoyClean Graffiti Remover®, Soyclean Formula 3 and Soy Release 3.

Blackline Masters

ACTIVITY SHEET 1 EXPLORE

Name

THE GROWTH OF SYNTHETIC CHEMICALS (part 1)

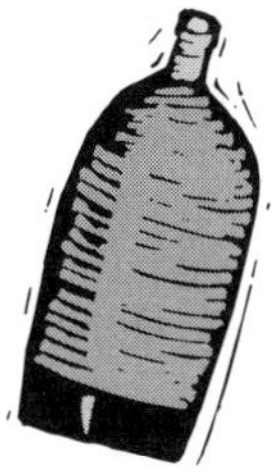

Make a line graph based on the information presented in the chart below. Then answer the questions.

Approximate Amount of Synthetic Chemicals Produced in the United States (in Tons)

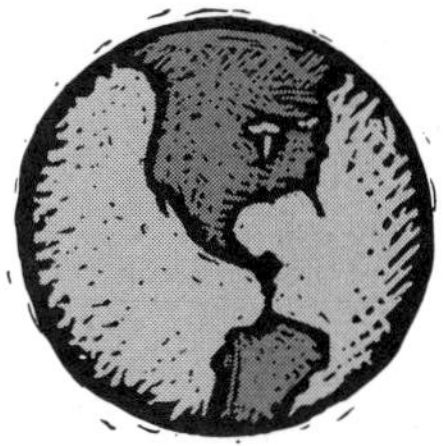

1945	8 million	1970	47 million
1950	8 million	1975	70 million
1955	12 million	1980	100 million
1960	18 million	1985	102 million
1965	45 million		

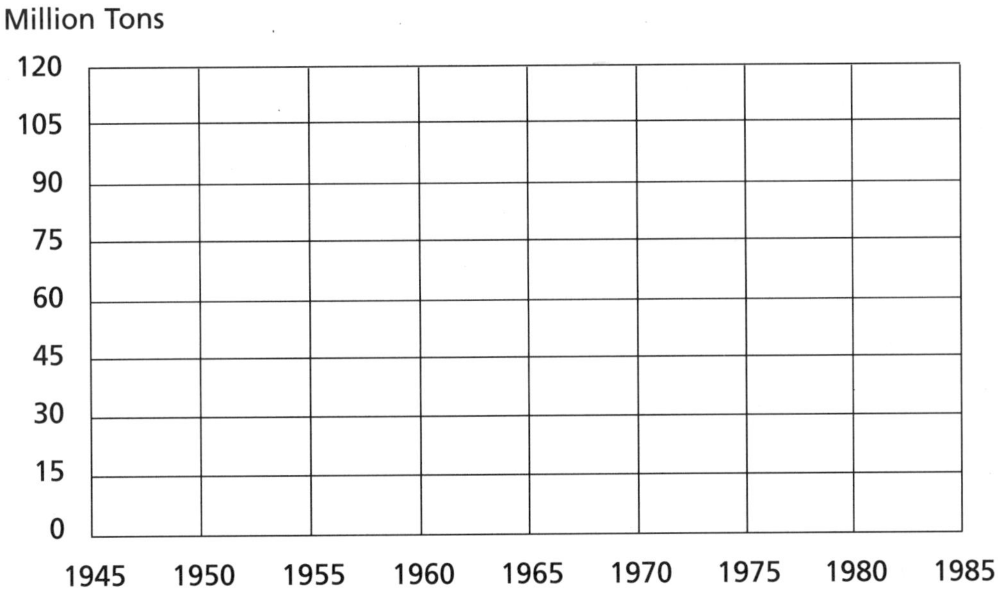

1. For every ton of synthetic chemicals produced in 1945, about how much more was produced in 1980? Circle the correct answer.

Twice as much Five times as much Twelve times as much

© The Tides Center/E2: Environment & Education

ACTIVITY SHEET

EXPLORE 1

Name ______________________________

THE GROWTH OF SYNTHETIC CHEMICALS (part 2)

2. From 1950 to 1980, which decade had the smallest increase in production of synthetic chemicals?

____

____

3. The line graph you drew stops at 1985. Look at what happened between 1980 and 1985. If that same trend were to continue, about how many million tons of synthetic chemicals would be produced in 2010?

____

____

____

____

____

____

____

Explain your reasoning.

____

____

____

____

____

____

____

____

© The Tides Center/E2: Environment & Education

ACTIVITY SHEET

EXPLORE

Name ______________________________

HAZARDS OF CHEMICAL PRODUCTS (part 1)

Study the picture and find as many possible sources of chemical pollution or hazards as you can. Circle each one, and put a number inside each circle. Then use the lines to list the numbers and explain how each circled hazard might affect humans, animals, and/or the environment.

____

____

____

____

____

____

© The Tides Center/E2: Environment & Education

ACTIVITY SHEET

EXPLORE

Name

HAZARDS OF CHEMICAL PRODUCTS (part 2)

© The Tides Center/E2: Environment & Education

ACTIVITY SHEET

Name

READING PRODUCT LABELS (part 1)

The warnings below are from actual product labels. The first label is for a product that removes mildew from certain surfaces. The second label is for a dish-washing detergent. Study the labels and then answer the questions.

PRECAUTIONARY STATEMENTS: Hazardous to humans and domestic animals.

WARNING: CAUSES EYE AND SKIN IRRITATION. Do not get in eyes or on clothing. Vapors may irritate. Use only in well-ventilated areas. Open windows and turn on fans before use. If the vapors bother you, leave the room while the product is working. Avoid prolonged breathing of vapors. Not recommended for use by persons with heart conditions or chronic respiratory problems such as asthma, emphysema, or obstructive lung disease. Harmful if swallowed. For sensitive skin or prolonged use, wear gloves. **STATEMENT OF PRACTICAL TREATMENT: EYE CONTACT:** Remove contact lenses and rinse eyes thoroughly with water for 15 minutes. **IF SWALLOWED:** Drink a glass of water and call physician. **IF BREATHING IS AFFECTED:** Get fresh air immediately. **SKIN CONTACT:** Remove contaminated clothing and wash affected area. **PHYSICAL AND CHEMICAL HAZARDS:** This product contains bleach. Do not use or mix with other household chemicals such as ammonia, toilet bowl cleaners, rust removers, vinegar, or acid. To do so will release hazardous gases. **STORAGE AND DISPOSAL:** Store in a cool place. Do not reuse empty container. Wrap in a plastic bag and discard.

CAUTION: May cause burns to skin, eyes, and mucous membranes. In case of external contact, flush with water for 15 minutes and get immediate medical attention. If swallowed, give a glassful of water or milk and call a physician. **KEEP OUT OF REACH OF CHILDREN.**

1. Which label indicates a more hazardous product?

© The Tides Center/E2: Environment & Education

Name ______________________________

READING PRODUCT LABELS (part 2)

2. For the first product, the mildew remover, what problems might occur in a small, windowless room?

3. For the second product, the dish-washing detergent, what should you do if you get some in your eyes?

4. What alternative products could you substitute for the mildew remover?

5. What alternative products could you substitute for the dishwashing detergent?

© The Tides Center/E2: Environment & Education

ACTIVITY SHEET 4 EXPLORE

Name ____

RECORDING DATA

Use the chart below to record data from the pest tolerance test. Read the statements in column 1. In column 2, make a mark for each person in the class who responds to that statement. Count the marks and write the total in column 3. Then answer the questions.

Pest Tolerance Test

Statement	Tally	Total
I would use a chemical pesticide if I found . . . • ants on the windowsill		
• ants crawling out of cracks in the wall		
• ants crawling on the floor		
• ants on my desk		
• ants inside my backpack or book bag		
• an ant crawling up my leg		
• an ant in my hair		
• ants crawling on the food in my lunch bag		
I would not use a chemical pesticide in any of these situations		

Without using pesticides, how can you . . .

- keep ants from crawling on a windowsill? ____

- keep ants from crawling out of cracks in a wall? ____

- keep ants out of the food you bring for lunch? ____

© The Tides Center/E2: Environment & Education

ACTIVITY SHEET

ANALYZE 5

Name ______________________________

CHEMICAL PRODUCTS IN THE COMMUNITY (part 1)

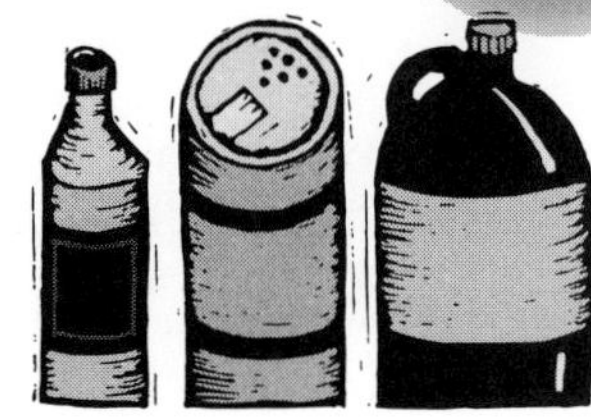

Use this sheet to record information as you listen to the presentation by the resource person from your community.

Pesticides, Insecticides, and Herbicides

Product/ Material	**Use**	**Health and Environmental Risks**	**Alternatives** (if presented)

© The Tides Center/E2: Environment & Education

ACTIVITY SHEET 5 ANALYZE

Name ______________________________

CHEMICAL PRODUCTS IN THE COMMUNITY (part 2)

Cleaning

Product/ Material	Use	Health and Environmental Risks	Alternatives

Maintenance

Product/ Material	Use	Health and Environmental Risks	Alternatives

© The Tides Center/E2: Environment & Education

ACTIVITY SHEET

ANALYZE

Name

SCHOOL CHEMICAL PRODUCTS USE (part 1)

On the notecards below, record ideas and information about hazardous materials used at your school. Use a different card for each type of material.

Material
What is it?

Why used?

Where used?

Hazards?

Site observations:

Are there other ways to deal with the problem or address the need?

Material
What is it?

Why used?

Where used?

Hazards?

Site observations:

Are there other ways to deal with the problem or address the need?

© The Tides Center/E2: Environment & Education

ACTIVITY SHEET

ANALYZE

Name

SCHOOL CHEMICAL PRODUCTS USE (part 2)

Material
What is it?

Why used?

Where used?

Hazards?

Site observations:

Are there other ways to deal with the problem or address the need?

Material
What is it?

Why used?

Where used?

Hazards?

Site observations:

Are there other ways to deal with the problem or address the need?

© The Tides Center/E2: Environment & Education

ACTIVITY SHEET

7 ANALYZE

Name ______________________________

CHEMICAL USE AUDIT PLAN (part 1)

Action Group: ______________________________

Action Group Focus: ______________________________

Member	Task	Questions	How to Get Information	Due Date
	Take notes at meetings.			

© The Tides Center/E2: Environment & Education

ACTIVITY SHEET

Name ____________________

CHEMICAL USE AUDIT PLAN (part 2)

Member	Task	Questions	How to Get Information	Due Date

© The Tides Center/E2: Environment & Education

ACTIVITY SHEET

ANALYZE

Name ______________________________

CHEMICAL USE AUDIT FORM

Use the chart to record the information you gather during your assessment of chemical use.

Product	**Site**	**Cost**	**Hazards/Label Warning**	**Additional Information** (intended use, frequency, etc.)

© The Tides Center/E2: Environment & Education

ACTIVITY SHEET 9

ANALYZE

Name ______________________________

CHEMICAL USE SUMMARY

Complete the table below to summarize your findings.

Product/Use	Pros	Cons

© The Tides Center/E2: Environment & Education

10 ANALYZE

Name ______________________________

ALTERNATIVES TO HARMFUL CHEMICALS (part 1)

Use the notecards below to organize your ideas about alternatives to harmful chemicals. Fill in a different card for each product.

Product

Alternative Product

Modifications to School Environment

Behavior/Attitude Change

Product

Alternative Product

Modifications to School Environment

Behavior/Attitude Change

Product

Alternative Product

Modifications to School Environment

Behavior/Attitude Change

© The Tides Center/E2: Environment & Education

ACTIVITY SHEET

10 ANALYZE

Name

ALTERNATIVES TO HARMFUL CHEMICALS (part 2)

Product

Alternative Product

Modifications to School Environment

Behavior/Attitude Change

Product

Alternative Product

Modifications to School Environment

Behavior/Attitude Change

Product

Alternative Product

Modifications to School Environment

Behavior/Attitude Change

© The Tides Center/E2: Environment & Education

Name ______________________________

COSTS/BENEFITS ANALYSIS OF ALTERNATIVES

Complete the table below to record your ideas.

Product/Use	Alternative	Costs	Benefits

© The Tides Center/E2: Environment & Education

Name ______________________________

PRESENTATION PLAN (part 1)

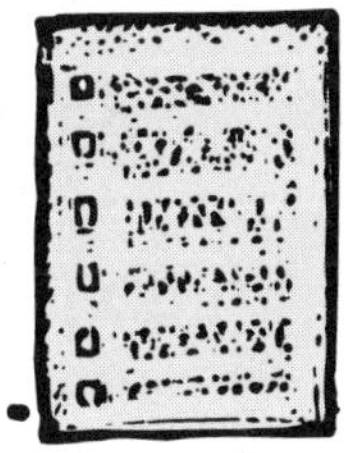

Use this sheet to organize the presentation your Action Group will give to the class when you recommend your alternatives.

Alternative ____________________ Current ____________________

Costs ______________________________

____

Benefits ______________________________

____

Implementation plan ______________________________

____

Alternative ____________________ Current ____________________

Costs ______________________________

____

Benefits ______________________________

____

Implementation plan ______________________________

____

Alternative ____________________ Current ____________________

Costs ______________________________

____

Benefits ______________________________

____

Implementation plan ______________________________

____

© The Tides Center/E2: Environment & Education

Name ______________________________

PRESENTATION PLAN (part 2)

List of Visuals You Will Use to Support Your Recommendations

Description	Source	Person Responsible

Use the space to sketch out a table, chart, or graph that you want to use to show some of your information.

© The Tides Center/E2: Environment & Education

Name ______________________________

RATING SHEET

Fill in the following rating sheet for each presentation.

Group ______________________________

Plan ______________________________

Costs

Expensive • • • Inexpensive

Health and Environmental Benefits

Low • • • High

Long-Term Effectiveness

Low • • • High

Difficulty of Implementing

Low • • • High

Cooperation Incentives

Low • • • High

Effectiveness of Presentation

Low • • • High

Additional Factors to Consider

____

____

____

Priority

Low • • • High

© The Tides Center/E2: Environment & Education

ACTIVITY SHEET 14 TAKE ACTION

Name ____________________

PROPOSAL CHECKLIST (part 1)

Use this checklist to plan and monitor tasks that may need to be done in order to complete your proposal. You can add to the list as needed.

Tasks	Person Responsible	Due Date
1. TITLE ☐ Cover illustration ☐ Proposal statement		
2. WRITE THE INTRODUCTORY PARAGRAPH. ☐ Explain the project. ☐ Briefly describe audit findings.		
3. WRITE RECOMMENDATIONS. ☐ Outline each plan. ☐ Highlight the benefits. ☐ Specify the costs. ☐ Suggest a step-by-step plan for implementation. ☐ List ideas for increasing awareness, changing attitudes, and encouraging support. ☐ Pinpoint projected savings.		

© The Tides Center/E2: Environment & Education

ACTIVITY SHEET 14 TAKE ACTION

Name

PROPOSAL CHECKLIST (part 2)

Tasks	Person Responsible	Due Date
4. PRESENT RESEARCH FINDINGS. ☐ Prepare graphs, if needed. ☐ Design tables or charts. ☐ Prepare illustrations, photographs, or other art work.		
5. WRITE CLOSING STATEMENT. ☐ Summarize briefly what steps you have recommended and why.		
6. MAKE A BIBLIOGRAPHY. ☐ List the resources that you used to generate your recommendations.		

© The Tides Center/E2: Environment & Education

ACTIVITY SHEET 15 TAKE ACTION

Name ______________________________

IMPLEMENTATION TRACKING SHEET

Use this tracking sheet to evaluate the status of each of your recommendations.

Proposal	Results to Date	Ideas for Follow-Up

Additional Ideas for Increasing Implementation

____

____

____

____

© The Tides Center/E2: Environment & Education

Name ____________________

CONTENT QUIZ (part 1)

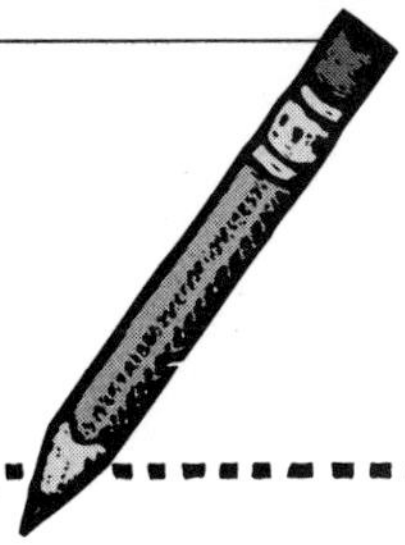

Circle the correct answer for each question. Some questions may have more than one correct answer.

1. The development of synthetic chemicals has decreased since the 1970s.

   a. true **b.** false

2. Furniture may be treated with chemicals that help repel insects and stains.

   a. true **b.** false

3. Chemicals absorbed into the ground can

   a. contaminate soil
   b. pollute the water supply
   c. both of the above
   d. none of the above

4. Toxic chemicals used today may cause harm to future generations.

   a. true **b.** false

5. A chemical that is combustible can

   a. pollute ground water
   b. irritate skin and eyes
   c. easily catch on fire
   d. harm future generations

6. It is safe to throw away containers that once had chemicals in them if they are completely empty.

   a. true **b.** false

7. Which of the following are used in labeling to indicate levels of toxicity?

   a. Danger, Hazardous, Poison, Inert
   b. Danger, Warning, Caution, Poison
   c. Hazardous, Caution, Poison, Toxic
   d. Poison, Toxic, Pollutant, Hazard

8. By law, all product ingredients must be listed on the label.

   a. true **b.** false

9. All chemicals in household products have been tested for toxicity.

   a. true **b.** false

© The Tides Center/E2: Environment & Education

Name ______________________________

CONTENT QUIZ (part 2)

10. By law, health risks that may be caused by toxic chemicals must be stated in product labels.

    a. true **b.** false

11. Herbicides are used to control

    a. plant-eating rodents

    b. slugs and snails

    c. weeds

    d. wildlife that is resistant to pesticides

12. Integrated Pest Management combines methods of controlling pests that offer alternatives to the use of chemical pesticides.

    a. true **b.** false

© The Tides Center/E2: Environment & Education

Name ____

STUDENT SURVEY

There are no correct answers to the questions below. Answer "yes" or "no."

	YES	NO	
1.	☐	☐	I know how to dispose of hazardous chemicals in my community.
2.	☐	☐	I know what hazardous chemicals are used in my home.
3.	☐	☐	I encourage my family to find alternatives to using hazardous chemicals.
4.	☐	☐	I know that hazardous chemicals are properly stored in my home.
5.	☐	☐	I try to find alternatives to using hazardous chemicals.
6.	☐	☐	I would rather live with pests than with pesticides.
7.	☐	☐	I know that we do not use toxic household products to clean our home.
8.	☐	☐	I know that we do not use herbicides on our garden.
9.	☐	☐	My family tries to find alternatives to using insecticides on the garden or on pets.
10.	☐	☐	I know that the best way to avoid the hazards of toxic chemicals is to find alternatives to their use.

© The Tides Center/E2: Environment & Education

Name ______________________________ Date ______________

Action Group ______________________________

STUDENT SELF-EVALUATION FORM

Evaluate your contributions to Action Group and class activities.

Poor	Average	Good	Excellent	
				Actively participated in Action Group/class discussions.
				Demonstrated a clear understanding of issues.
				Took responsibility for research and other tasks.
				Provided suggestions and solutions to problems.

How could I have improved my participation in this project?

____

____

What skills did I improve upon? ______________________________

____

Which need further improvement? ______________________________

____

As a result of participating in this project:

What new knowledge have you gained? Have you shared knowledge with others at school?

____

How has your opinion about environmental issues/problems changed? ______________________________

____

Have you made any changes in your daily life and/or home? ______________________________

____

____

© The Tides Center/E2: Environment & Education

Action Group ______________________________

Date ______________________ Time ______________________

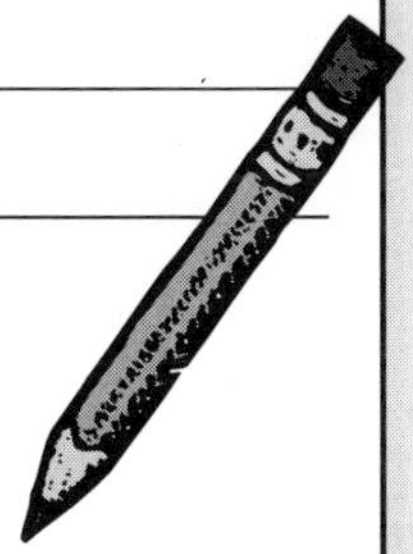

ACTION GROUP EVALUATION FORM

Behavioral Functions	Participants' Names				
1. Initiating discussion					
2. Sharing information					
3. Seeking information					
4. Giving opinions					
5. Seeking opinions					
6. Evaluating information					
7. Clarifying information					
8. Dramatizing points					
9. Coordinating tasks					
10. Finding compromises					
11. Suggesting procedures					
12. Recording information					
13. Finding middle ground					
14. Relieving tension					
15. Finding norms					
16. Withdrawing from group					
17. Blocking discussion					
18. Seeking recognition					
19. Horsing around					
20. Advocating for group members					
21. Dominating discussion					
22. Criticizing others					

© The Tides Center/E2: Environment & Education

PROGRAM EVALUATION FORM (part 1)

Teachers: In an effort to continually upgrade this module, we are asking for constructive criticism and suggestions for improvement. Please complete an evaluation form for each module and return to:

E2: Environment & Education
P.O. Box 20515
Boulder, CO 80308-3515
303/442-3339 Fax: 303/442-6633
email: e2ee@earthlink.net

Name ______________________________

School ______________________________

School address ______________________________

Grade(s) taught ______________________________

---

1. Please check the module(s) that you used:

   ☐ Energy Conservation ☐ Water Conservation ☐ Waste Reduction
   ☐ Food Choices ☐ Habitat and Biodiversity ☐ Chemicals: Choosing Wisely

2. What did you like most about the module(s)?

   ____
   ____
   ____

3. What did you like least about the module(s)?

   ____
   ____
   ____

4. Please comment on the following aspects of the modules:

   Readability
   ____
   ____

   Instructions
   ____
   ____

   Organization
   ____
   ____

   Graphics and Layout
   ____

© The Tides Center/E2: Environment & Education

Name ____

PROGRAM EVALUATION FORM (part 2)

Readability

____

____

Instructions

____

____

Organization

____

____

____

5. Rate how the module affected your students' environmental awareness.

	Not at all				Very much so
Awareness of issues	1	2	3	4	5
Knowledge of issues	1	2	3	4	5
Sense of responsibility	1	2	3	4	5
Sense of ability to affect change	1	2	3	4	5

6. Rate the overall involvement of the following groups in conducting and implementing environmental activities at your school.

	Not at all				Enthusiastically
Students	1	2	3	4	5
Maintenance Staff	1	2	3	4	5
School Administrators	1	2	3	4	5
Parents	1	2	3	4	5
Teachers	1	2	3	4	5
Local Community Organizations	1	2	3	4	5
Local Businesses	1	2	3	4	5
Others ___________	1	2	3	4	5

© The Tides Center/E2: Environment & Education